CATASTROPHE THEORY AND BIFURCATION

Catastrophe Theory and Bifurcation

APPLICATIONS TO URBAN AND REGIONAL SYSTEMS

A. G. WILSON

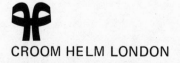

CROOM HELM LONDON

UNIVERSITY OF CALIFORNIA PRESS
BERKELEY AND LOS ANGELES

© 1981 A.G. Wilson
Croom Helm Ltd, 2-10 St John's Road, London SW11

British Library Cataloguing in Publication Data

Wilson, Alan Geoffrey
    Catastrophe theory and bifurcation. -
    (Croom Helm series in geography and
    environment).
    1. Sociology, Urban - Mathematical models
    2. Catastrophes (Mathematics)
    3. Bifurcation theory
    I. Theory
    301.36'01'5147   HT153
    ISBN 0-7099-2702-9

University of California Press
Berkeley and Los Angeles

Library of Congress Cataloging in Publication Data

Wilson, Alan Geoffrey.
    Catastrophe theory and bifurcation.

    Bibliography: p.301
    Includes index.
    1. Geography—Mathematics. 2. Catastrophes (Mathe-
matics) 3. Bifurcation theory. I. Title.
G70.23.W54   910'.01'51472   80-6059
ISBN 0-520-04370-7

Printed and bound in Great Britain
by Billing and Sons Limited
Guildford, London, Oxford, Worcester

PREFACE

I first became interested in catastrophe theory in late
1975 after the publication in English of Rene Thom's
Structural stability and morphogenesis.   At that time, I was
also interested in shopping centre dynamics and was first
beginning to work with differential equations for this purpose.
Later, I was to connect this to bifurcation theory in its more
general sense.   It has turned out that my interests have been
shared by a number of other geographers and planners, both as
practitioners, but more widely, as readers.

The purposes of this book, therefore, are: first, to offer
an elementary account of catastrophe theory and bifurcation
theory for the benefit of social scientists interested in
cities; secondly, to review applications in the field to date;
and thirdly, by tackling a number of problems in depth (which,
it is hoped, typify much of geographical theory), indicate
where some of the current research frontiers are.   There is
also a chapter devoted to a review of applications in other
disciplines.

It is hoped that the book will be of interest to
geographers, regional scientists and planners and to members of
other disciplines interested in cities (economists, engineers
and, to an increasing extent, applied mathematicians and
operational researchers for example).

I gave a short course of lectures on these topics, first
in the University of Pennsylvania in September 1978, and then
in the University of Leeds later in the same year.   I have
also given lectures and seminars in other universities on the
same topics.   I am grateful to colleagues and members of those

universities for much helpful discussion and comment. I should
particularly mention John Beaumont, Martin Clarke, Paul Keys
and Huw Williams, who read and commented on various chapters as
they were written. I am particularly grateful to John Amson
of the Mathematical Institute of the University of St. Andrews
who read a draft of the whole manuscript and made a number of
very helpful comments. As usual, I alone am responsible for
the final shortcomings. I am grateful to Pamela Talbot for
producing the typescript and to Gordon Bryant and John Dixon
for drawing the figures. I am especially grateful to
Pamela Talbot, Tim Hadwin and John Dixon for all their work in
producing the final camera-ready copy.

ACKNOWLEDGEMENTS

Figures 1.4, 1.15 & 7.4 are reproduced from Figures 1 & 2 in
A.G. Wilson (1976) Catastrophe theory and urban modelling:
an application to modal choice, *Environment and Planning A,
8*, 351-6.
Figures 4.10, 4.11, 4.12 & 4.13 are based on Figures 1, 2, 4 &
5 in T. Poston and A.G. Wilson (1977) Facility size vs.
distance travelled: urban services and the fold catastrophe,
Environment and Planning A, 9, 681-6.
Figures 1.6, 1.7, 1.8, 5.9 & 5.16 are based on Figures 1, 4 & 5
in B. Harris and A.G. Wilson (1978) Equilibrium values and
dynamics of attractiveness terms in production-constrained
spatial interaction models, *Environment and Planning A, 10*,
371-88.
Figure 6.3 is based on Figure 6.3 in A.G. Wilson (1978) Towards
models of the evolution and geneis of urban structure, in
R.L. Martin, N.J. Thrift and R.J. Bennett (Eds.) *Towards
the dynamic analysis of spatial systems*, Pion, London, 79-90.
All reproduced by permission of Pion Ltd., London.

Figures 4.2, 4.3 & 4.4 are based on Figures 5, 6 & 7 in
J.C. Amson (1974) Equilibrium and catastrophic modes of urban
growth, in E.L. Cripps (Ed.) *Space-time concepts in urban
and regional models*, Pion, London, 108-28.
Figures 1.1, 1.13 & 1.12 are based on Figures 2, 3 & 7 in
J.C. Amson (1975) Catastrophe theory: a contribution to
the study of urban problems? *Environment and Planning B, 2*,
177-221.
Figures 1.5, 7.8, 7.9, 7.10 & 7.11 are based on Figures 3, 4, 5
& 6 in J.H. Blase (1979) Hysteresis and catastrophe theory:
empirical identification in transport modelling, *Environment
and Planning A, 11*, 678-88.
Figures 4.22, 4.23 & 4.24 are based on Figures 3, 4 & 5 in
G. Papageorgiou (1980) On sudden urban growth, *Environment
and Planning A, 12*, 1035-50.
Figures 7.12, 7.13 & 7.14 are based on Figures 4, 7 & 8 in
J.L. Deneubourg, A. de Palma and D. Kahn (1979) Dynamic
models of competition between transport modes, *Environment
and Planning A, 11*, 665-73.
All reproduced by permission of Pion Ltd., London and the
respective authors.

Figures 2.16, 2.17 & 2.20 are based on Figures 7, 8 & 28 in
J. Maynard Smith (1974) *Models in ecology*, Cambridge
University Press, Cambridge.
Reproduced by permission of Cambridge University Press.

Figures 8.3, 8.4, 8.5 & 8.6 are based on Figures 1, 2, 3 & 4 in R.M. May (1976) Simple mathematical models with very complicated dynamics, *Nature, 261,* 459-67.
Reproduced by permission of Macmillan Journals, London.

Figures 8.7 & 8.8 are based on Figures 1 & 4 in E.C. Zeeman (1974) Primary and secondary waves in developmental biology, in *Lectures on mathematics in the life sciences,* American Mathematical Society, Rhode Island, Vol. 7, 69-161.
Reproduced by permission of The American Mathematical Society.

Figures 8.14, 8.15, 8.16 & 8.17 are based on Figures 3, 4, 5 & 6 in H.L. Varian (1979) Catastrophe theory and the business cycle, *Economic Inquiry, 17,* 14-28.
Reproduced by permission of The Western Economic Association, U.S.A.

Figure 8.18 is based on Figure 3 in F. Boon and A. de Palma (1978) Boolean formalism and urban development, Mimeo.
Reproduced by permission of the authors.

CONTENTS

Page

PREFACE   v

ACKNOWLEDGEMENTS   vii

CONTENTS   xi

LIST OF FIGURES   xvii

LIST OF TABLES   xxi

CHAPTER 1.   A LAY GUIDE TO THE MATHEMATICS OF CATASTROPHE
THEORY   1

1.1   The nature of catastrophe theory   1

1.2   A preliminary outline of some examples   6

     1.2.1   A remark on scale   6
     1.2.2   The Zeeman catastrophe machine   7
     1.2.3   The micro scale: modal choice and the cusp
           catastrophe   8
     1.2.4   The meso scale: spatial structure and the
           fold catastrophe   10
     1.2.5   The macro scale: city growth and the cusp
           catastrophe   12

1.3   An informal review of the mathematical concepts   14

     1.3.1   An outline of the basic concepts   14
     1.3.2   The mathematics of the elementary catastrophe   18
     1.3.3   Sets on the behaviour manifolds, and delay
           conventions   20
     1.3.4   The fold catastrophe   22
     1.3.5   The cusp catastrophe   25
     1.3.6   The remaining elementary and higher order
           catastrophes   28

CHAPTER 2.   DIFFERENTIAL EQUATIONS AND BIFURCATION   33

2.1   Differential equations and non-gradient systems   33

2.2   Dynamical systems and solutions to differential
equations: a sketch of basic concepts   34

     2.2.1   A preliminary note on types of graphical
           presentation   34
     2.2.2   Equilibrium points and trajectory sketching   37
     2.2.3   Static models embedded in dynamic frameworks   38
     2.2.4   Basic types of trajectory   39
     2.2.5   Bifurcation   42

2.3   Examples   43

    2.3.1   Introduction   43
    2.3.2   Growth equations   43
    2.3.3   Competition 1: the Lotka-Volterra equations 46
    2.3.4   Competition 2: fixed resources   49
    2.3.5   Logistic growth equations for interacting
             populations   54
    2.3.6   Further extensions: linked subsystems and
             fluctuations   55

CHAPTER 3.   APPLICATIONS OF DYNAMICAL SYSTEMS THEORY:
                 A SURVEY OF APPROACHES   57

3.1   Introduction   57

3.2   Differential equations and catastrophe theory   58

3.3   Relative speeds of change: variables, parameters and
    constants; system description   59

3.4   Levels of approach   60

3.5   Qualitative vs. quantitative; inductive vs.
    deductive   63

3.6   A new focus for planning applications of models   64

NOTE   65

CHAPTER 4.   MACRO-SCALE APPLICATIONS   67

4.1   Introduction   67

4.2   Amson (1974): catastrophic modes of urban growth   69

4.3   Casti and Swain (1975) 1: central place theory   74

4.4   Casti and Swain (1975) 2: property prices   75

4.5   Poston and Wilson (1976): another approach to
    centre size   78

4.6   Mees (1975): the revival of cities in medieval
    Europe   82

4.7   Isard (1977): strategic elements of a theory of
    major structural change   82

4.8   Wagstaff (1978): settlement pattern evolution   84

4.9   Dendrinos (1977): slums in urban settings   88

4.10   Papageorgiou (1980): sudden urban growth   89

4.11   Concluding comments   92

CHAPTER 5. BIFURCATION AT THE MESO-SCALE I:
            COMPARATIVE STATICS OF URBAN SPATIAL
            STRUCTURE   93

5.1 Introduction   93

5.2 The examples to be used   94

     5.2.1 The urban retail structure model   94
     5.2.2 Other models with a similar structure   98
     5.2.3 Residential structure   98
     5.2.4 Interacting fields: the Lowry model   99
     5.2.5 Disaggregation   101
     5.2.6 Composite attractiveness factors   105
     5.2.7 Theoretical foundations of the models:
            mathematical programming   108
     5.2.8 Summary of the position reached   108

5.3 Equations for dynamical analysis   110

     5.3.1 Introduction   110
     5.3.2 Retail model differential equations   110
     5.3.3 Difference equations   114
     5.3.4 Residential location   115
     5.3.5 The Lowry model   116
     5.3.6 Disaggregated models   116

5.4 Equilibrium point analysis   116

     5.4.1 Introduction   116
     5.4.2 Retail model equilibrium point theory   118
     5.4.3 Summary of results on retailing: towards a
            theory of structural evolution   141
     5.4.4 Residential location   144
     5.4.5 The Lowry model   148
     5.4.6 Disaggregated models   149
     5.4.7 Ecological analysis   150

NOTES   153

CHAPTER 6. BIFURCATION AT THE MESO-SCALE II: THE
            DYNAMICS OF URBAN SPATIAL STRUCTURE   155

6.1 Introduction   155

6.2 Disequilibrium, fluctuations and bifurcation   155

     6.2.1 Introduction   155
     6.2.2 Order from fluctuations: the Brussels
            school   156
     6.2.3 The use of kinetic equations   168
     6.2.4 Concluding comments   170

6.3 Control theoretic formulations   171

     6.3.1 Introduction   171
     6.3.2 A control problem in shopping centre
            location   172
     6.3.3 Other possible applications   173

6.4   Integrated approaches: towards a new intra-urban
        central place theory   173

        6.4.1   Introduction: the bases of central place
                theory   173
        6.4.2   An alternative model representation for
                central place theory   178
        6.4.3   An example of an SIA model to be used as
                the basis for intra-urban central place
                theory   179
        6.4.4   Urban dynamics and the SIA model of central
                place theory   186
        6.4.5   The evolution of urban structure   191
        6.4.6   Comparisons with traditional central place
                theory   196

6.5   Some possibilities for further research   196

        6.5.1   Introduction: development or evolution?   196
        6.5.2   Further research on models in the
                development mode   197
        6.5.3   An example of the evolution of new
                structures   199
        6.5.4   Concluding comments   201

NOTES   202

CHAPTER 7.   MICRO-SCALE APPLICATIONS   203

7.1   Introduction   203

7.2   Some general considerations   203

7.3   Hysteresis and modal choice   206

        7.3.1   Hysteresis, catastrophe theory and modal
                choice   206
        7.3.2   A mechanism for hysteresis   208
        7.3.3   Empirical evidence   210

7.4   Dynamic modal choice models and bifurcation   210

        7.4.1   Some principles   210
        7.4.2   Model 1: attractiveness proportional to
                speed   212
        7.4.3   Model 2: addition of psychological factors   215

7.5   The speed-flow relationship and the fold catastrophe   218

        7.5.1   The empirical results   218
        7.5.2   A behavioural model and the fold catastrophe   219

7.6   Concluding comments   222

Page

CHAPTER 8.   APPLICATIONS IN OTHER DISCIPLINES AND SOME
              NEW RESULTS FOR URBAN SYSTEMS   225

8.1   Introduction   225

8.2   Physical chemistry   226

       8.2.1   Kinetic equations: interacting mixtures   226
       8.2.2   Dissipative structures: order from
               fluctuations   229

8.3   Biology   232

       8.3.1   Introduction   232
       8.3.2   Developmental biology   232
       8.3.3   Evolutionary biology   234

8.4   Ecology   236

       8.4.1   Population dynamics: differential equations   236
       8.4.2   Population dynamics: difference equations   240
       8.4.3   Travelling waves in ecological models   246

8.5   Economics   248

       8.5.1   Introduction   248
       8.5.2   Resource management   249
       8.5.3   Business cycles   253

8.6   Applications to cities   256

       8.6.1   Introduction and review   256
       8.6.2   Urban models and Boolean algebra: analogues
               of genetic switching   256
       8.6.3   Difference equations and shopping centres   260

CHAPTER 9.   CONCLUDING COMMENTS   267

APPENDIX 1.   MATHEMATICAL PROGRAMMING FORMULATIONS OF THE
              MAIN MODELS   271

A1.1   Introduction   271

A1.2   Entropy maximising and the shopping model   271

A1.3   Consumers' surplus   275

A1.4   Embedding: optimum centre size and location   278

A1.5   Accessibility maximising: Leonardi's formulation   282

A1.6   Random utility theory and group surplus   283

A1.7   Application to residential location models   285

A1.8   Mathematical programming versions of the Lowry model   285

APPENDIX 2.   SOME ALTERNATIVE LAGRANGIAN FORMULATIONS   291

APPENDIX 3.   THE DERIVATIVES OF S_{ij} 295

APPENDIX 4.   THE SHOPPING TRIP FLOW DERIVATIVES FOR THE
              DISAGGREGATED MODEL   297

REFERENCES AND BIBLIOGRAPHY   301

INDEX   315

LIST OF FIGURES

		Page
1.1	The cusp surface	4
1.2	The Zeeman catastrophe machine	7
1.3	Modal choice and the cusp catastrophe	9
1.4	Sections parallel to the (x,u_2)-plane of the modal choice cusp catastrophe	9
1.5	Empirical evidence of hysteresis	9
1.6	Intersecting revenue and cost curves for retailing against centre size	11
1.7	Non-intersecting retail revenue and cost curves	11
1.8	Retailing centre size and the fold catastrophe	11
1.9	Central place rank and the cusp catastrophe	13
1.10	Commuting mappings as diffeomorphisms	16
1.11	Plots of $f(x) = x^3 + \alpha x$	19
1.12	Illustration of catastrophe, bifurcation and conflict sets for the cusp	21
1.13	The fold catastrophe	23
1.14	Plots of objective function curves for typical u values	25
1.15	The control manifold and plots of the objective function for typical (u_1,u_2) values for the cusp catastrophe	26
1.16	Local maximum and local minimum created by the constraint $x > 0$	31
2.1	Trajectories in state space	34
2.2	Trajectory in behaviour manifold in two dimensions	35
2.3	State variable plotted against intermediate variable	35
2.4	A state variable vs. a parameter, exhibiting bifurcation	36
2.5	A state variable plotted against time	36
2.6	A state variable plotted against its time derivative	36
2.7	State equilibrium points in state space	40
2.8	Time plots representing progressions to a stable equilibrium point	40
2.9	Unstable equilibrium points in state space	40
2.10	Time plots representing divergence from an unstable equilibrium point	40
2.11	A saddle point	41
2.12	Periodic trajectories in state space	41
2.13	Plots of $x = ke^{ut}$ for varying u	44
2.14	Logistic growth	45

2.15   Plots of the solutions of $x = \varepsilon(D - x)x^n$ 46
2.16   Stable equilibrium points for the Lotka-Volterra
       equations   47
2.17   An unstable configuration for the Lotka-Volterra
       equations   48
2.18   An alternative Lotka-Volterra configuration with the
       predator eliminated   48
2.19   A fold catastrophe for parameter change in the
       Lotka-Volterra system   49
2.20   Alternative state-space configuration for the
       competition-for-resources (C-F-R) model   50
2.21   Different 'solution regions' on the control manifold
       for the C-F-R model   51
2.22   C-F-R model plots of $\dot{x}_1 = 0(A_1)$, $\dot{x}_2 = 0(A_2)$ 52
2.23   C-F-R model: A_1, A_2 non-intersecting   53
2.24   C-F-R model: A_1, A_2 intersecting   53

4.1   Amson's third law and the fold catastrophe   72
4.2   Amson's fourth law and the cusp catastrophe   73
4.3   Amson's fourth law: the control manifold   73
4.4   Amson's fourth law: typical paths in the control
       manifold   73
4.5   Casti and Swain: order of central places and the
       cusp catastrophe   75
4.6   Casti and Swain: urban property prices and the
       cusp catastrophe   76
4.7   Sections of the (u_1,u_2) control manifold for the
       butterfly catastrophe: $u_1 > 0$ 77
4.8   Sections of the (u_3,u_4) control manifold for the
       butterfly catastrophe: $u_1 < 0$ 77
4.9   Casti and Swain: a representation of the butterfly
       catastrophe manifold   78
4.10   Poston and Wilson: two components of utility for
       shopping centres   79
4.11   Poston and Wilson: total utility   79
4.12   Poston and Wilson: different equilibrium points for
       changing 'ease of travel'   80
4.13   Poston and Wilson: shopping centre size and the
       fold catastrophe, with additional state   81
4.14   Mees: city growth and the butterfly catastrophe -
       sections of control manifold and sample trajectories   83
4.15   The cusp catastrophe as total welfare   84
4.16   Wagstaff: area of study, space coordinate as
       distance along a curve   85
4.17   Wagstaff: the cusp catastrophe   86
4.18   Wagstaff: possible trajectories on the control
       manifold   86
4.19   Wagstaff: the potential function for changing
       values of the control variables   86
4.20   Wagstaff: settlement densities - taken as potential
       function   87
4.21   Dendrinos: urban slums and the mushroom catastrophe   88

4.22	Papageorgiou: urban and rural utilities - with economies of scale	90
4.23	Papageorgiou: effect of technological change on urban size	90
4.24	Urban change as represented by a section of the cusp catastrophe	91
5.1	Flows to shopping centres	96
5.2	Flows to workplaces	100
5.3	Interactions of two fields	101
5.4	The main urban subsystem	109
5.5	Plot of $(\alpha - 1)/2\alpha$	121
5.6	Retailing: flow-centre size curves	122
5.7	Points of inflexion on flow size curves	124
5.8	Retailing: revenue-size curves	124
5.9	Revenue-size curves	125
5.10	Stability considerations - revenue-cost-size curves	125
5.11	Revenue curve with varying cost lines	127
5.12	Effect of increasing α on revenue curve	127
5.13	Effect of decreasing β on revenue curve	127
5.14	$k_j^{crit} - \beta_j^{crit}$ curve	128
5.15	$k_j^{crit} - \alpha_j^{crit}$ curve	128
5.16	Retailing centre size and the fold catastrophe, k as control variable	129
5.17	Retailing centre size and the fold catastrophe, α as control variable	129
5.18	Retailing centre size and the fold catastrophe, β as control variable	129
5.19	Flow vs. centre size	133
5.20	Plot of revenue surves, less one flow, relative to cost	134
5.21	Flow, revenue less one flow: two cases and the critical case	134
5.22	Critical case: alternative presentation	134
5.23	Cost line transformed to cost curve	136
5.24	Criticality and near-criticality with cost curve	138
6.1	The structure of the Lowry model	180
6.2	Capacity and cost functions vs. retail centre size	199
6.3	Graphical representation of linear programming model of evolution of shopping centre size	200
7.1	Distribution of behaviour types	205
7.2	Cusp model for behaviour types	205
7.3	Modal choice and the cusp catastrophe	206
7.4	A section of the modal choice cusp surface	207
7.5	Modal probability vs. cost difference: step function	208
7.6	Modal probability vs. cost difference: modal bias normally distributed	209
7.7	Modal choice probability for the whole population	210
7.8	Petrol price and weekend traffic levels	211

7.9   Saturday traffic vs. petrol price   211
7.10   Modal choice hysteresis loop   211
7.11   Fitted lines for sections of hysteresis loop, with
        confidence limits   211
7.12   Modal choice bifurcation diagrams: (a) x_1 vs. D,
        (b) x_2 vs. D   215
7.13   Conditions for solution of Equation (7.44)   217
7.14   Bifurcation diagrams for the cases (a) $\Theta_2 > (\alpha_1\alpha_2)^{\frac{1}{2}}$,
        (b) $\Theta_2 < (\alpha_1\alpha_2)^{\frac{1}{2}}$ 217
7.15   Traffic: speed-density and flow-density curves   219
7.16   Traffic: speed-flow curve   219
7.17   Driver utility components vs. speed and flow   220
7.18   Total driver utility vs. speed   220
7.19   Speed-flow curve and the fold catastrophe   221
7.20   Utility functions vs. speed at densities above a
        critical value   221

8.1   Chemical concentration vs. ratio of reaction rates   227
8.2   Chemical concentration vs. parameter bifurcations   231
8.3   Logistic difference equation x_{t+1} vs. x_t 242
8.4   Logistic difference equation: stability of
        equilibrium solution   243
8.5   Logistic difference equation: stability of one-
        period oscillation   244
8.6   Logistic difference equation: bifurcation points   245
8.7   Environmental gradient and travelling frontier   246
8.8   Cusp catastrophe explanation of travelling function   247
8.9   Natural growth and harvesting functions for a fish
        population, and their difference   250
8.10   Yield as a function of effort   250
8.11   Alternative natural growth functions   250
8.12   Differences of natural growth and harvesting
        functions for alternative growth functions   251
8.13   Yield vs. effort and a fold catastrophe   252
8.14   Investment and savings function vs. national income   254
8.15   Plots of $\dot{y} = 0$, $\dot{k} = 0$, in state space   254
8.16   Alternative savings function, with investment
        functions   255
8.17   Depression and the cusp catastrophe   255
8.18   Boolean state table for an urban system   259
8.19   Retail centre size: W_{jt+1} vs. W_{jt} for a linear
        difference equation II model   261

A1.1   Hierarchical relationships between micro, meso and
        macro state descriptions   273
A1.2   Change in consumers' surplus   276
A1.3   Total consumers' surplus, axes reversed   277

LIST OF TABLES

		Page
1.1	The seven elementary catastrophes	29
6.1	Model parameters for SIA central place system	187
6.2	Model parameters and assumptions associated with them	188

CHAPTER 1

A LAY GUIDE TO THE MATHEMATICS OF CATASTROPHE THEORY

1.1 The nature of catastrophe theory

Most mathematical models represent a state of some system
of interest as an equilibrium point which is a function of some
parameters or independent variables.   When the parameters
change slowly and smoothly, then the position of the equilibrium points also changes slowly and smoothly. Catastrophe
theory, on the other hand, is concerned with sudden and discrete changes in system state variables resulting from a slow,
smooth and small change in one or more parameters.

The theory arises from the mathematical work of René Thom,
whose famous book, *Structural stability and morphogenesis*, was
published in French in 1972 and English in 1975.   The title
demonstrates another of the major concerns of catastrophe
theory: with the stability of *forms* and the creation of *forms*.
Each of the concepts of stability and creation imply also a
general concern with *dynamical analysis*.

The underlying mathematics of catastrophe theory relies on
the techniques of differential topology and this takes it
beyond the grasp of those with a relatively elementary education in old-style calculus.   This is certainly true of the
deepest theorems on which the theory is based.   The main
objectives of this book, therefore, are to attempt to convey
some of the main ideas in an informal and necessarily
unrigorous way and to outline the potential of their application mainly in urban and regional studies but also in a number
of other disciplines.   If the basis of some of the ideas, and
the nature of the applied problems they can contribute to, are
understood, then this may encourage, first, some new
elementary work from this perspective alone;   secondly, workers

in urban and regional studies (and other disciplines in a
similar position) to take up differential topology; and
thirdly, mathematicians with the appropriate skills to become
interested in this kind of applied work.

The title of the book refers both to catastrophe theory
and bifurcation more generally. This is because the second
kind of branching, or jump, behaviour is relevant to systems of
a more general type than the first, and we shall also develop
the argument in this direction. Indeed, it is likely to turn
out that this kind of bifurcation is the most important in much
applied work. Most of the early illustrations arise from
catastrophe theory directly, however. This limits systems of
interest to so-called *gradient systems* which arise from the
minimisation of some objective function and associated dynamics
(or maximisation of its negative, so maximisation is covered as
well). For example, let $\underline{x}$ be a set of state variables describing some system - the dependent variables to be predicted in a
model - and let $\underline{u}$ be a set of parameters - or independent variables. The $\underline{u}$-variables are frequently referred to as control
variables which is appropriate for many contexts. Then, in a
gradient system, the equilibrium position is determined by

$$\underset{(\underline{x},\underline{u})}{\text{Min}} = f(\underline{x},\underline{u}) \tag{1.1}$$

for some function, f. The dynamics of the process is given by

$$\underline{\dot{x}} = -\frac{\partial f}{\partial x} = -\text{grad } \underline{f} \tag{1.2}$$

and the minimum of f, of course, occurs when

$$\text{grad } \underline{f} = 0 \tag{1.3}$$

The appearance of the gradient of the potential function f
explains the name of this type of system. Although it seems
to be restrictive, many systems can in fact be described in
this way even if it is not always apparent at first sight
(*cf.* Wilson and Macgill, 1979; Macgill and Wilson, 1979).

Mathematics of catastrophe theory

The solution to equations (1.3) give the equilibrium point which minimises the potential function in (1.1) and, as $\underline{u}$ varies, determine a surface in the space $(\underline{x},\underline{u})$. This is a surface representing *possible* equilibrium states of the system. If, for example, there is a single state variable and two control variables - x_1 and (u_1,u_2) respectively - then this will be a surface in the three dimensional space (x_1,u_1,u_2). This will be illustrated shortly with an example in Figure 1.1.

The conventional mathematics of equations (1.1)-(1.3) is well known. As noted earlier, for a smooth, slow and small change in one or more of the $\underline{u}$ variables, a corresponding smooth change in the state variables $\underline{x}$ can be anticipated. It can be seen intuitively that for this to occur, the surface in $(\underline{x},\underline{u})$ space of equilibrium solutions has to be itself smooth and not *folded* in any way. It has long been recognised that when, for a given $\underline{u}$, there are multiple solutions for $\underline{x}$, then something more complicated can occur. The essence of Thom's work is the classification of these complications and the proof that, in a number of cases, they fall into a small group of basic types.

The solutions of (1.1) (or equivalently (1.3)) are the stationary points of the function f or, more precisely, of a family of functions of $\underline{x}$, parametrised by $\underline{u}$. Stationary points are often maxima or minima, which are well-known types and are distinguished by, in the single state variable case, the second derivative of f being negative or positive respectively. (In the multi-state-variable case, the corresponding result is that the Hessian matrix of second derivatives is negative- or positive-definite, respectively.) When stationary points are not maxima or minima, the second derivative is zero or the Hessian matrix is singular. Such equilibrium points are known as *singularities* and it is at and near such points that unusual system behaviour can be observed and predicted. What Thom, and others, have done is to classify the kinds of singularity which can occur. It has

3

been shown that, for a number of control variables in the vector $\underline{u}$ up to or equal to 4, the types of singularity, in a topological sense, are relatively few.   For example, in the case of a single state variable and two control variables, the surface of equilibrium points around a singularity *must be* topologically equivalent to the now-famous cusp surface.   This is illustrated in Figure 1.1.

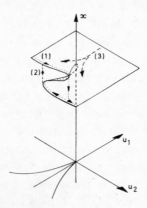

Figure 1.1 The cusp surface

We can illustrate the possibilities of catastrophe theory using this figure.   The surface of possible equilibrium values describes all possible states of the system.   A particular *behaviour* of the system is a trajectory on the surface.   The study of such surfaces for particular systems, therefore, allows us to investigate possible types of behaviour, and Thom's theorem can be used to restrict the possibilities because we know, as noted above, that the surface must *in a topological sense* be of the form shown in the figure.   The italicised qualification, however, is an important one in practice, and should be emphasised even at this early stage. In effect, 'in a topological sense' means that the surface of possible equilibrium values for a system can be forced into the form of Figure 1.1 after some smooth transformation of the variables where necessary.   This is known as a standard, or canonical form. The achievement of the appropriate transformation in applied work is often likely to be a very difficult task, as we shall see later, though insights can often be gained without it being carried through explicitly.

Three types of behaviour which we are *not* accustomed to expect are shown in sample trajectories on Figure 1.1. They are:

(1)   a sudden *jump* (or catastrophe);
(2)   *hysteresis* - a reverse path to some point not being the same as the original;   and
(3)   *divergence* - a small difference in approach towards, in this case a cusp point, leads the system to the upper or lower surface and hence to a very different state.

It can easily be seen that the jump behaviour arises from a path in the u-plane which leads the system to 'fall' from the upper surface to the lower one at a *fold* - or vice versa.

It can also be seen, thinking back to the way in which the surface was constructed, that folds, and hence jump behaviour, arises because in some regions of u-space, there are *multiple* equilibrium solutions for x.   In the particular case of Figure 1.1, there is a region in the central part of the diagram where there appear to be three possible solutions for x. It turns out that the upper and lower surfaces represent stable minima (and hence are observable) while the central part of the fold represents maxima and hence unstable (and unobservable) states.   If this folded region is projected vertically down-wards onto the u-plane, we obtained the familiar cusp-shaped section of that plane.   This contained the set of values of u which are in some sense *critical*: outside the shaded region, the system only has one state available to it;   inside there are two possible states and hence possible *conflict*;   as the boundary of the critical region is crossed, jumps can take place. We will see later how this preliminary analysis can be connected to various concepts of stability. We will also see explicitly that, as noted above, the function f is singular at *critical* parameter values and that geometrically this can be identified with folds in the equilibrium surface. In the critical region, where there are multiple states, some rule has to be assumed, or discovered, about which state the system actually adopts. This involves a so-called *delay convention*, which we will also pursue later.

Thus, the main new idea arising from catastrophe theory as discussed so far is that it is worthwhile being alert to the possibility of unusual types of system behaviour, and that the techniques of the theory can be used to formulate models to handle such behaviour. In the next section, we begin a preliminary treatment of some examples and follow this with a slightly more formal treatment of the basic concepts.

1.2 A preliminary outline of some examples
1.2.1 A remark on scale

As usual in urban and regional modelling, it is important to distinguish the different scales at which models are to be developed. The usual distinction of micro-, meso- and macro-scales is useful here as elsewhere. At the micro scale, we are concerned with individual behaviour of some kind, and decisions, possibly under constraints (and with the resulting behaviour being in some cases determined more by the constraints than any notion of choice). At the meso scale, we deal with urban and regional spatial structure, and this is the scale which is at the heart of geographical analysis (even though it has to be intimately connected to other scales). At the macro scale, the city or region is characterised by a small number of aggregate variables.

It is particularly important to make these distinctions for urban and regional studies, as most of the current applications of catastrophe theory are at micro or macro scales, probably because the theory is essentially designed to handle only a small number of variables (and hence 'simple' systems - *cf*. Weaver's classification, discussed by Wilson, 1977-B). As usual, we find that some of the most difficult problems are at the meso scale and that, if used in a particular but hitherto unusual way, catastrophe theory can help with these problems.

To wet appetites, an urban example is presented below at each of the scales. First, however, a short description is offered of the Zeeman catastrophe machine, as this offers the opportunity for any sceptics to build such a machine and to convince themselves that systems do change states in a fast and discrete way.

1.2.2 The Zeeman catastrophe machine

This machine can be simply constructed and is shown in Figure 1.2. A is a fixed point - a nail - on a board. A disc is pivoted at O, and B is a fixed point on the disc. AB is an elastic band. BP is a second elastic band, and P is a free end which can be held in the hand and moved around. The state of

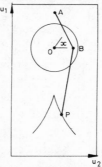

Figure 1.2 The Zeeman catastrophe machine

the system is the position of the disc, measured say by an angle, x, and the control variables are u_1 and u_2 - Cartesian coordinates measuring the position of P. Thus u_1 and u_2 are the independent variables and x is the dependent variable. Experiment with such a machine shows that the disc jumps when P is taken across the second of the cusp curves. This illustrates both jumps and hysteresis in an obvious way. What is happening is that the position of the disc is determined by the minimum (potential) energy of the two elastic bands, and the possible behaviours are described by the cusp catastrophe in a direct way. For a detailed account, see Zeeman (1977-A) or Poston and Stewart (1978).

Chapter 1

1.2.3. The micro scale: modal choice and the cusp catastrophe

At the micro scale, we are usually concerned with decisions which are essentially discrete, and choice of transport mode is an obvious example of this. This causes some difficulties as the state variables we have considered so far are continuous. We handled this, possibly glossing over some mathematical difficulties in doing so, by defining a state variable x where $x < 0$ denotes choice of mode 1 and $x > 0$, choice of mode 2. We then take as control variables u_1, a habit factor, and u_2, a term proportional to the difference (say) of costs of the two modes. Then, in principle, we have a situation described by the cusp surface, as depicted in Figure 1.3.

We can ease the analysis by taking sections of the curve for different fixed values of u_1. That is, we let the plane u_1 = constant cut the cusp surface for a series of values of the constant. This produces two cases, as shown in Figure 1.4. We have a plot therefore of x against u_2 for a series of values of u_1. The two cases represent (a) a plot for negative u_1 where the plane intersects the folded part of the curve, and (b) for positive u_1, where the intersection is 'above' the fold. Sample trajectories are shown on the figure. The main point of interest relative to more conventional models is that in case (a), there is a clear hysteresis effect. If a mode loses passengers through an increase in fares say, then a corresponding drop in fares would not necessarily get the passengers back.

This kind of hysteresis effect has also been explored from another perspective by Goodwin (1977), and has been investigated empirically by Blase (1979) with some interesting results as shown in Figure 1.5.

This example should, of course, be treated with considerable caution. There is no inherent reason, as was emphasised in a general context earlier, why the three variables of this model should be related in exactly the way

Mathematics of catastrophe theory

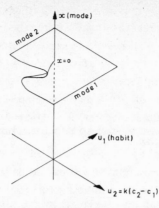

Figure 1.3 Modal choice and the cusp catastrophe

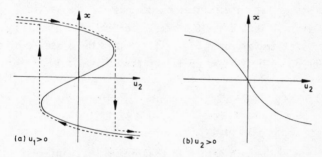

Figure 1.4 Sections parallel to the (x, u_2) - plane of the modal choice cusp catastrophe

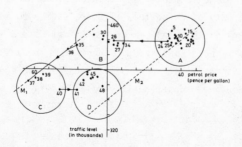

Figure 1.5 Empirical evidence of hysteresis

shown without transforming the variables, possibly in a non-linear way. However, it does illustrate that thinking of possible models in terms of catastrophe theory can introduce new ideas, in this case the hysteresis phenomenon, which can then be introduced into other existing models. It also raises new empirical questions for investigation.

1.2.4 The meso scale: spatial structure and the fold catastrophe

This example will be pursued in greater detail later, as it has many interesting features, so only a sketch of the central idea will be given here. Consider the supply of some service, say shopping to fix ideas, measured by $\{W_j\}$. This is related to some given zoning system in the usual way. Then, such a system can be described by demands for the service, spatially distributed as $\{O_i\}$, and a pattern of usage given by $\{S_{ij}\}$. Let travel costs be $\{c_{ij}\}$. Then, the system can be represented by a standard spatial interaction model

$$S_{ij} = A_i O_i W_j^\alpha e^{-\beta c_{ij}} \qquad (1.4)$$

where

$$A_i = 1/\sum_j W_j^\alpha e^{-\beta c_{ij}} \qquad (1.5)$$

(this is described in detail in Chapter 5). An interesting recent development is to add mechanisms to this model by which facility sizes, W_j, can be determined endogenously (and the detailed argument for this is presented by Harris and Wilson, 1978). One way of doing this illustrates the role of catastrophe theory in the analysis of spatial structure. This involves assuming a 'balancing mechanism' whereby the W_j-pattern adjusts so that 'revenue' (or usage) and supply balance through a proportionality relationship for each j:

$$D_j = kW_j \qquad (1.6)$$

(where D_j is the revenue in zone j and is equal to $\sum_j S_{ij}$).

The model given by equations (1.4) and (1.5) can also be used to give D_j as a function of W_j and the corresponding geometrical curve is called the revenue curve. Then the solutions for W_j can be seen as arising from the intersections of the revenue curve, the model-based D_j - W_j curve, and the straight line (in effect, a cost curve) given by equation (1.6). This situation is depicted in Figure 1.6. It turns out that the stable points are the origin (which means no facility in that zone) and W_j^{opt} as shown. We should add that this shape

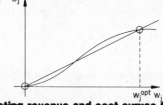

Figure 1.6 Intersecting revenue and cost curves for retailing against centre size

of curve relies on the condition $\alpha > 1$. It is immediately clear that the existence or otherwise of facilities in zone j depends on the value of k (which is obviously the slope of the straight line) and the particular D_j - W_j curve for that zone. If k increases from the value shown in Figure 1.6, we get the situation shown in Figure 1.7, and the non-zero W_j^{opt} has disappeared.

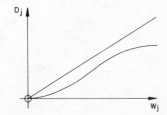

Figure 1.7 Non-intersecting retail revenue and cost curves

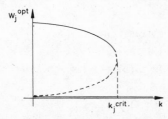

Figure 1.8 Retailing centre size and the fold catastrophe

This information can be presented as on Figure 1.8: a plot of W_j^{opt} against a variable k. This plot, as we shall see later, is characteristic of the fold catastrophe: at some critical value of k (which depends on the zone, because the revenue curve is specific to the zone, and is hence labelled k_j^{crit}), facilities appear or disappear according to the direction of change. The importance of this point at this stage is the indication it gives of new methods for modelling the evolution of spatial structure. This is pursued in more detail below.

1.2.5 The macro scale: city growth and the cusp catastrophe

Casti and Swain (1975) offer a simple illustration of the use of the cusp catastrophe to examine the 'size' of a city (measured as the order of that city in some central place system) as a function of two control variables which are taken as population and per capita income. Their interpretation of the dynamics of this system is shown in Figure 1.9. This shows that the city order, m, increases steadily with population, unless the trajectory crosses the second side of the critical region at point II. If the trajectory is in the reverse direction, then the jump downwards will come at the point I, thus illustrating the hysteresis effect. In central place theory terms, this represents a mechanism by which the threshold which has to be achieved for a particular level is higher for the appearance of a function (point II) than it is for its disappearance (point I).

The income factor plays the same sort of role as the habit factor in the modal split example above: for sufficiently high incomes, the transitions between different levels will be smooth. As we will see below, it is acting as the *splitting factor* while population is acting as the *normal factor*.

This example also illustrates divergence. If a trajectory approaches the cusp point (which would represent a situation in which income was decreasing from above the cusp

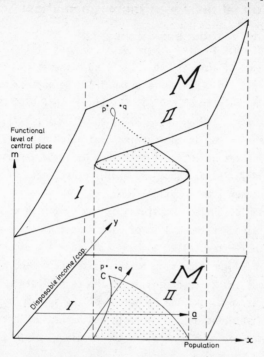

Figure 1.9 Central place rank and the cusp catastrophe

and the population figure was coincidentally at this critical
position), then a small difference in the value of the population factor determines which sheet of the surface the
system moves on to: a move to the left of the cusp point (from
above) takes the system on to the lower sheet, while a pass to
the right takes it onto the upper sheet.   While there is evidence in the literature of the differing threshold values for
growth and decline, there is no evidence as yet of this sort of
divergence phenomenon.   This is another interesting example of
how the use of catastrophe theory can raise new empirical
questions.

The problem, of course, with this sort of macro analysis
is that there are many alternative ways in which control variables can be chosen.   This is a difficulty which will be much
discussed in later sections.

1.3 An informal review of the mathematical concepts

1.3.1 An outline of the basic concepts

We now attempt to give a more detailed, but still informal account of some of the underlying mathematics of catastrophe theory. This review is drawn from the work of many authors, but notably Amson (1975), Chillingworth (1975, 1976), Poston and Stewart (1978), Thom (1975) and Zeeman (1977). It will be convenient, as in the earlier sketch, mostly to discuss a particular system of interest described by state variables $\underline{x}$ and control variables $\underline{u}$, though much of the argument, of course, is a general mathematical one.

The possible equilibrium states of the system as we have seen, are obtained as the minima of some function f, and they form a surface - often called a *manifold* - in n + k dimensional space, where n is the dimension of the state space (the number of $\underline{x}$-variables) and k the dimension of the control state (the number of $\underline{u}$-variables). The behaviour of the system is described by trajectories on the manifold and hence our interest in the geometry, or more generally, the topology, of this surface.

Thom showed that the shape of the manifold, in a sense we will explore further below, depends on the number of control variables, k (known as the co-dimension of a family of functions parametrised by these variables) and on the co-rank of any singularities associated with the potential function, f. Roughly speaking, the co-rank measures the degree of degeneracy of the worst kind of singularity which can occur in the particular family of functions. (For functions of one variable, the degree of degeneracy is n, where all derivatives up to the nth order vanish.)

For Thom's results on the so-called elementary catastrophes, the co-dimension must not be greater than four and the co-rank not greater than two. This second condition is represented as the number of state variables needed to construct the degenerate singularities - a number equal to the co-rank - and in effect means that the catastrophe theory, by means of canonical

transformations of variables can handle a large number of
state variables as though there were only one or two, though
only up to four control variables. (We will show below how the
remaining state variables are treated.) Thus, we will usually
be concerned with potential functions like $f(x_1,u_1,u_2)$ or, at
worst, $f(x_1,x_2,u_1,u_2,u_3,u_4)$.

As noted earlier, the mathematical analysis can only be
carried through at a proper level of depth if the perspective
of differential topology is used. This will not be seriously
attempted here, though occasionally definitions will be introduced to enable connections to be made between this more
traditional and primitive argument and the catastrophe theory
literature. A topological viewpoint essentially involves the
idea of functions as mappings from points in one topological
space to those in another. This allows analysis to be conducted for arbitrary dimensions and avoids too much reliance
being placed on geometrical intuition (though that is a useful
commodity in other ways in this context).

Thom's basic concern is with the notion of structural
stability of models. The initial argument has a powerful
intuitive appeal: many of the features we are trying to model
of real-world systems are inherently stable. When we build
models, we use analytical functions which are unlikely to be
an exact representation. Therefore, if the model is going to
work, its essential features should be stable against small
perturbations in the functions used. Thom was searching for
measures of this kind of *structural stability* of functions used
in models. The mathematical results which emerge from this
investigation are powerful and surprising. We can now begin
to outline these results by extending the ideas sketched above.

The equilibrium states of the system of interest are
determined by the stationary points (or, to be more precise,
some of them - the minima) of the potential function governing
the system. Morse and Thom and other mathematicians have
discovered that for a wide class of functions (which cover many

of the functions to be used in model building) the
singularities at the critical equilibrium points fall into a
small number of types. It is this classification of these
families of functions which lies at the heart of catastrophe
theory. The essence of the idea of one function being of the
same type as another in this respect turns on the possibility
of smoothly transforming one into the other by a change of
variables. Amson (1975) says that such functions are of the
same *diffeotype*. This arises from the concept of a diffeomorphism in topology which relates two topological spaces.
Let two functions f and g be mappings from topological spaces P
to Q and R to S respectively. Then if there is a diffeomorphism between P and R, say p, and another between Q and S, say
q, and the diagram showing all the mappings involved (see
Figure 1.10) commutes, then the functions are of the same
diffeotype.

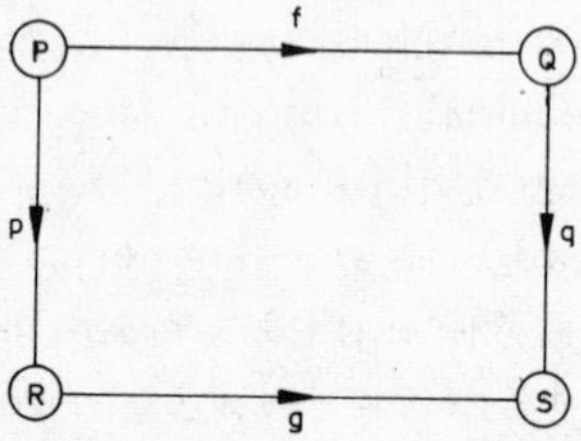

Figure 1.10 Commuting mappings as diffeomorphisms

This brief excursion into topology shows the difficulty in
thinking in this way for those of us not trained in the subject.
Perhaps, however, it will suffice to recognise in some intuitive sense that functions can be of the same topological type,
and that this sameness involves being able to transform one
into the other through a smooth change of variables. It is
important to recall in all subsequent deliberations, however,
that such a change of variables will usually be necessary to
get the function into standard form and that it will be surprising if applications involve using functions in their
canonical form directly (though occasionally this does happen).
    Recall, first, that we are interested in the surface of
equilibrium points which are the minima of the potential

function.   It is at the singular points of the function where
such equilibrium points appear and disappear.   Bearing this in
mind, it turns out that most functions are of a relatively
limited number of types:

(1)   they have no singularities - and these are much the most
      common;

(2)   they have singularities of a regular kind - called Morse
      singularities - which are essentially representable as
      $x_1^2 + x_2^2 + \ldots - x_r^2 - x_{r+1}^2 - \ldots$.   These points are well-
      defined maxima or minima, or saddle points, in an
      appropriate number of dimensions;   or

(3)   they have *degenerate* singularities, and in many cases
      these are of a limited number of types depending only on
      the co-rank and co-dimension of the singularity.

Catastrophe theory arises out of the third case.   In the
second case, the equilibrium points are found by setting the
derivatives of the potential function to zero (as in equation
(1.3)), but the Hessian matrix of second derivatives is non-
singular - and in this sense the points are non-degenerate.
When the Hessian matrix is singular, then we have the third
case.   If it is of rank r (and there are n state variables)
then n - r is the co-rank of the singularity.   The limited
number of types of singularity is defined in relation to the
notion of families of functions, and the number of parameters
needed to define a family.   The number of parameters is the
co-dimension of the family, and if it is finite, then the number of types of degenerate singularity is finite.   Otherwise
the family is of infinite co-dimension, and we do not pursue
these any further here (though they lead to the concept of
generalised catastrophes and will doubtless be of interest and
importance at some future time - see Thom, 1975).

     If the co-dimension is less than or equal to four, then
the number of types of singularity is very small, and indeed
Thom's seven elementary catastrophes refer to this case.   This
also lets us see the importance of identifying the control

variables for our system of interest: if their number can be
restricted to four or less, then Thom's theorems can be used to
describe the kinds of singularity which can occur, and this in
turn provides a description of the *possible* types of trajectory
of system behaviour on the equilibrium surface.

1.3.2 The mathematics of the elementary catastrophe

The next step is to progress towards an intuitive understanding of Thom's classification theorem using the concept of
Taylor expansions. Unfortunately, the theorem cannot possibly
be proved with the mathematical equipment we have available to
us here. However, we can examine the types of singularity in
relation to canonical (or standard) forms of functions and
exploit the theorem's result that other functions of the same
co-rank and co-dimension can then be transformed (locally, in
the neighbourhood of a point) into the same form. The canonical forms are polynomials which (following Amson, 1975), in
general, for a single state variable x, take the form

$$f(x) = x^m + u_1 x^{m-2} + u_2 x^{m-3} + \ldots + u_{m-2} x \qquad (1.7)$$

The first term, in this case x^m, captures the degeneracy
and type of the singularity. If all the u-variables are zero,
this can be considered as the lowest order non-zero term in a
Taylor expansion. As the u-variables then vary from zero
values, the right hand side of equation (1.7) *approximates* the
Taylor expansion of a whole family of functions. Thom's
theorem is essentially saying that all other families of functions with the same number of parameters have singularities of
the same type as the canonical, truncated Taylor expansion,
form. This form is said to represent a universal unfolding of
singularities of this type.

Thom's theorem says that for m up to six (that is, up to
four control variables) this "models all functions of that
co-dimension and the structure of the singularities, in this
neighbourhood of the function in its canonical form can be used
as a 'model' for the singularities of all the functions of this
type".

The notion of 'unfolding' can also be expressed in another way, based on the concept of 'structural stability' which, as we noted earlier, provides another route into catastrophe theory. Consider the function

$$f(x) = x^3 + ux \qquad (1.8)$$

which is a special case of equation (1.7) with m = 3. This is plotted in Figure 1.11 for the cases u < 0, u = 0 and u > 0.

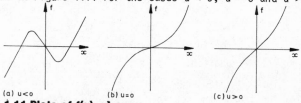

(a) u < o   (b) u = o   (c) u > o

Figure 1.11 Plots of $f(x) = x^3 + ax$

When u = 0, f(x) = x^3, and this is *not* structurally stable in the sense that the addition (or subtraction) of a term ux, however small u, changes the shape of the curve in a basic way in the neighbourhood of the origin. The function f(x) in (1.8) when u ≠ 0, however, *is* structurally stable: it retains its shape under small perturbations. f(x) = x^3 is said to have a degenerate singularity at x = 0, and the addition of the term ux is the simplest way to make the function structurally stable. This is also said to be an unfolding of the denegerate singularity.

It is at this point that another important qualification must be *added*. As $(u_1, u_2 ...)$ vary the family of functions span a function space. The power of Thom's theorem relates to the simple classification of the topology of the singularities of a function with a certain number of control variables. But the theorem is only true locally - that is, in the polynomial case, in a neighbourhood of the degenerate singularity at x = 0. If the whole surface can be constructed explicitly, as for the canonical polynomial form, this is fine; if not, some caution has to be exercised in relation to the meaning of 'locally' in particular cases: this has caused some of the controversy in the applications of catastrophe theory.

19

The power of the theorem stems from the proof that there
is only a small number of such types - seven, as noted earlier
for the elementary catastrophes. We will use the most elementary, the fold and the cusp, to illustrate this argument below,
and present a summary of the remainder in a table. First,
however, we complete our introduction of general concepts.

1.3.3 Sets on the behaviour manifolds, and delay conventions

We can now complete our general discussion by returning to
the heart of the matter: catastrophes. We noted earlier that
catastrophes occur because of the existence of multiple minima
of the potential function. The behaviour manifold is defined
as the surface in ($\underline{x},\underline{u}$) space which contains the minima of the
potential function - the possible equilibrium states of the
system. We can usefully classify different possible types of
system behaviour by focusing on the *control manifold* - the
smaller-dimensional $\underline{u}$-space. For each point on the control
manifold, consider the point or points (if any) to which it
gives rise on the behaviour manifold. Then, we can identify
regions of the control manifold as follows. First, there is
that where those values of the control variables generate only
one equilibrium position, and the behaviour of the system is
then well-determined; secondly, there is the region (a comple-
mentary one by definition - or almost so) where there is more
than one solution (discounting for the time being any region
where there may be no solution). This is known as the
catastrophe set and it is not immediately clear which state the
system adopts - additional information must be supplied. The
bifurcation set is the set of points which separates the catastrophe set from the 'single solution' set; it is the critical
set of points at which a minimum disappears. It is at such
points that the system, if it is in the state which disappears,
must jump to another state - and hence branch (or bifurcate -
and hence the name of this set). These concepts are illustrated for a particular case in Figure 1.12, which is an
annotated version of Figure 1.1.

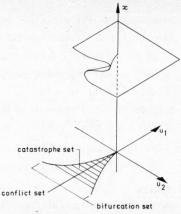

Figure 1.12 Illustration of catastrophe, bifurcation and conflict sets for the cusp

The behaviour of the system for control points within the catastrophe set is determined by a *delay convention*.   This is a rule which must be supplied to determine which of the multiple possibilities the system adopts.   The two most common are firstly, perfect delay, which means that the system stays in its original state until that state disappears as the trajectory leaves the bifurcation set;   and secondly, the Maxwell convention, which assumes that if more than one minimum is available, the system chooses the state which represents the lowest. These result in different kinds of behaviour, both of which are important in applications.   In the perfect delay case, jumps take place as the trajectory crosses the bifurcation line, as noted;   in the Maxwell case, the region of interest is the so-called *conflict set* defined as the points on the control manifold at which two or more minima take *equal* values.   This is also shown on Figure 1.12 for the cusp case. With perfect delay, system behaviour can be associated with the idea of thresholds which the system must cross before a change (as we saw with the central place theory example of Casti and Swain above);   in the case of the Maxwell convention, the conflict set can be seen as a travelling wave which is the basis of morphogenesis and another of Thom's original motivations in formulating the theory.   This last case turns out to be

21

particularly important where the control variables are taken as representing space and time (three space coordinates and one time coordinate). It is in this context that the travelling wave concept is important, and we will pursue this with examples later.

There is also the possibility of mixed conventions: the system could jump to an alternative state (when it is within the catastrophe set) according to some probability formula (which may be based on quite complicated notions).

1.3.4 The fold catastrophe

The simplest of Thom's elementary catastrophes is the *fold*. It is the universal unfolding of the singularities of x^3 and its potential function (which we examined briefly in a discussion of structural stability above) is

$$z = \frac{1}{3} x^3 + ux \qquad (1.9)$$

for a single state variable x and a single control variable u. This is the canonical form for a family of functions $f(x,u)$. We drop the usual subscripts, for convenience, in this simple case. It can easily be seen that this is a special case of the more general unfolding formula (1.7), in fact with m = 3. The possible equilibrium states of this system are those for which z is a minimum and we can find these by setting the derivative to zero:

$$\frac{dz}{dx} = x^2 + u = 0 \qquad (1.10)$$

and this has solutions

$$x = \pm(-u)^{\frac{1}{2}} \qquad (1.11)$$

and we note that the second derivative is

$$\frac{d^2z}{dx^2} = 2x \qquad (1.12)$$

Since this is positive for positive values of x, and negative for negative values, this shows that the minima occur for the

22

positive values and the maxima for negative values.
Equation (1.11) also shows that real roots only exist for
negative u.   This information is recorded on Figure 1.13.
This follows directly from Equation (1.10), which is that of a
parabola.   The top half has been shown as a solid curve,
because this represents the minima and the stable observable
states of the system, while the bottom half is dashed and
represents the maxima (which are unstable and unobservable).

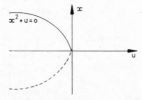

Figure 1.13 The fold catastrophe

We can now illustrate the general argument of the previous
subsection by this simple example.   The function z in
Equation (1.9) is a canonical representation of any function
with a singularity of co-rank 1 at the originand of
co-dimension 1.   In this case, since we have only one state
variable and one control variable, the whole picture of possible
equilibrium values - the singularities of z in the neighbourhood of the origin, can be represented in two dimensions as
shown on Figure 1.13.   The control manifold, the projection of
the (x,u) manifold onto the u-manifold is in this case simply
the horizontal axis.   There is no catastrophe set because
there are no points on the horizontal axis at which there are
two or more values of x for which z is a minimum, and the
bifurcation set is also very simple: it is the single point at
the origin, because here an observable minimum disappears.   It
is at this point therefore that jump behaviour can be observed:
if the system is in a state given by negative u and on a trajectory in which u is increasing, then as u passes through
zero, the stable minimum equilibrium state disappears and the
system will have to take up some other state not accounted for
by this diagram.   In fact, it is interesting in this context

to refer back to Figure 1.8 in one of the examples presented earlier (Section 1.2.4): we can now see this as essentially the fold catastrophe but with alternative states added, so that when the system passes through its equivalent of the origin, it jumps to a stable state at *its* origin.

Because there is no area of the control manifold which produces multivalued solutions, we cannot illustrate directly the concepts of delay conventions, conflict sets and so on. However, we have seen by reference back to Figure 1.8 that even in this case, states can be added in the particular application which do create multivaluedness. That example was concerned with the emergence or otherwise of spatial structure and obviously delay conventions and associated concepts of thresholds are very important. However, we will illustrate these notions using the cusp catastrophe below.

One other technique can be introduced at this stage which gives more insight into the workings of catastrophe theory. Plots such as Figure 1.13 give the equilibrium values of x and u but do not show what is happening to the function z in the neighbourhood of those values. This can be recorded on another form of diagram (originally due to Zeeman, 1977-A) as in Figure 1.14. For typical values of the control variable - in this case, u < 0, u = 0 and u > 0, we can plot z against the state variables, in this case x. In the u < 0 case, the minimum (occuring at a positive value of x) can easily be seen as can the way in which the plot of z against x changes as u increases from a negative value. At u = 0, the graph is an obvious limiting case: the maximum and minimum have fused to form a point of inflexion, while for u > 0, the stationary points have clearly disappeared. We will see a more complicated form of this diagram in relation to the cusp catastrophe below.

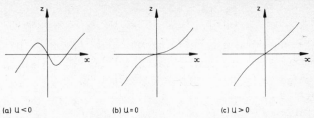

Figure 1.14 Plots of objective function curves for typical u values

1.3.5 The cusp catastrophe

The cusp catastrophe involves the universal unfolding of the singularities of the function x^4, of co-rank 1 and co-dimension 2, so that two parameters are needed to generate a full family of functions for the unfolding. The version of Equation (1.7), for this case is

$$z = \tfrac{1}{4}x^4 + \tfrac{1}{2}u_1x^2 + u_2x \qquad (1.13)$$

where we now add the subscripts to distinguish the two control variables but again omit the subscript for the state variable since there is only one. This is the canonical form for a family of functions $f(x,u_1,u_2)$. The stationary values of z are found by setting its derivative to zero:

$$\frac{dz}{dx} = x^3 + u_1x + u_2 = 0 \qquad (1.14)$$

We now have to recall our knowledge of the algebra of the cubic equation to discover the conditions under which different numbers of roots exist for different values of (u_1,u_2). Such an equation can have either one or three real roots. The condition for the existence of three real roots is

$$(-\tfrac{1}{3}u_1)^3 > (\tfrac{1}{2}u_2)^2 \qquad (1.15)$$

and, of course, this implies that

$$u_1 < 0 \qquad (1.16)$$

By squaring each side of Equation (1.15) we can see that the boundary of the region is defined by

25

$$4u_1^3 + 27u_2^2 = 0 \qquad\qquad (1.17)$$

This, as numerical experiment quickly shows, produces the cusp shaped curves on the control manifold - the (u_1,u_2) plane. This is plotted in Figure 1.15 together with a series of plots of z against x for various typical values of u_1 and u_2 (the cusp equivalent of Figure 1.14). This shows that outside the cusp shaped region, there is only one root and this is always a minimum. (This can also be seen from the fact that $z(\infty)$ and $z(-\infty)$ are both positive.)

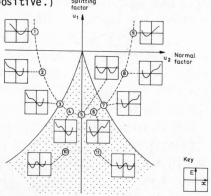

Figure 1.15   The control manifold and plots of the objective function for typical (u_1, u_2) values for the cusp catastrophe

Inside the region, there are three real roots and always one maximum (an unstable state) and two minima as can be checked by an examination of the second derivative of z. Thus the shaded region is the catastrophe set and the boundary of that region is the bifurcation set where a local minimum disappears. This can be seen happening at points 3 and 7 on Figure 1.15: the minimum which disappears merges into the local maximum to create a point of inflexion on the boundary. The u_1 axis, for $u_1 < 0$, represents the conflict set: where there are two minima of equal value as represented by point 5 on the figure.

The surface of equilibrium (x,u_1,u_2)-values is the famous folded surface first shown as Figure 1.1 and as Figure 1.12.

It can now easily be seen that the cusp shaped critical region on the control manifold is the projection of the folded part of this surface. We showed earlier, in the introduction, how the different kinds of catastrophic behaviour - jumps, hysteresis and divergence - can be exhibited in different possible trajectories on this surface. We can also see as sections of this surface, for example by taking a plane perpendicular to the control manifold and parallel to the u_2 axis generate for $u_1 < 0$, the kinds of s-shaped sections which were shown in Figure 1.4(a). Each side of such a section can be seen as representing a fold catastrophe, and this is a common phenomenon: higher order catastrophe surfaces are made up of sets of lower order catastrophes together with some distinctive new feature - in this case the cusp point. This is a particularly useful idea in relation to some of the very much higher order catastrophes where visual presentation in two or three dimensions is impossible.

The details of any system trajectories depend crucially on the delay convention which is operating. In the case of perfect delay, jumps will occur when the projection of the system trajectory on the control manifold crosses the second edge of the critical region. For example, in Figure 1.15 if the projected trajectory is represented by the dashed curve, then even though the trajectory crosses the bifurcation line at point 3, it will stay in the 'positive x' minimum until it reaches point 7 when that minimum disappears. If the same trajectory is then covered in reverse, then the jump will come at point 3, rather than 7, showing that it is the perfect delay convention which produces hysteresis effects. This is exactly the behaviour displayed by the Zeeman catastrophe machine of Section 1.2.2. This arises because the position of the disc is determined by the minimisation of a potential function - in fact the potential energy of the two elastic bands - and, because there are two control parameters only, Thom's theorem tells us that the dynamics arising from that function can be approximated at least locally by the cusp catastrophe in its canonical form.   For a full treatment of this analysis see either Zeeman (1977-A) or Poston and Stewart (1978).

If the Maxwell convention is operating, then it is the u_1 axis, the conflict set, which is of most interest and jumps will take place as the trajectory crosses that axis.

This is an appropriate point to introduce some useful terminology of Zeeman's (1977-A) in relation to control variables.   For this cusp case, he calls u_1 the 'splitting factor' and u_2 the 'normal factor'.   This is because it is the value of u_1 which determines whether a trajectory is in a region where the surface is folded.   With the coordinates chosen here, if $u_1 > 0$, then the surface is single valued, but it is two-valued within the region for which $u_1 < 0$.   In the case of the normal factor, u_2, x changes monotonically as u_2 changes, and continuously except for jumps at the bifurcation points.

1.3.6 The remaining elementary and higher order catastrophes

The seven elementary catastrophes are listed in Table 1.1. The table gives the number of state variables, the number of control variables and the potential function which gives the universal unfolding of that type of singularity.   Only the fold and cusp can be given a fully explicit geometrical treatment (as in the previous two subsections) because in other cases four or more dimensions would be needed for equivalent presentations.   However, it is possible to generate pictures by portraying two or three dimensional slices of higher order diagrams.   The full treatment for the remaining elementary catastrophes are, however, now available in many other sources which can be used for reference.   A particularly good detailed account is given by Amson (1975), but accounts also appear in most of the books listed at the beginning of this section.

The list of elementary catastrophes can be extended slightly by looking at *duals*.   These exist for three of the entries in Table 1.1: the cusp, the butterfly and the parabolic umbilic; the rest of the list are self dual.   Duals

Name	State variables	Control variables	Potential function
Fold	1	1	$\frac{1}{3}x_1^3 + u_1x_1$
Cusp	1	2	$\frac{1}{4}x_1^4 + \frac{1}{2}u_1x_1^2 + u_2x_1$
Swallow-tail	1	3	$\frac{1}{5}x_1^5 + \frac{1}{3}u_1x_1^3 + \frac{1}{2}u_2x_1^2 + u_3x_1$
Hyperbolic umbilic	2	3	$\frac{1}{3}x_1^3 + \frac{1}{3}x_2^3 + u_1x_1x_2 - u_2x_1 - u_3x_2$
Elliptic umbilic	2	3	$\frac{1}{3}x_1^3 - \frac{1}{2}x_1x_2^2 + \frac{1}{2}u_1(x_1^2 + x_2^2) - u_2x_1 - u_3x_2$
Butterfly	1	4	$\frac{1}{6}x_1^6 + \frac{1}{4}u_1x_1^4 + \frac{1}{3}u_2x_1^3 + \frac{1}{2}u_3x_1^2 + u_4x_1$
Parabolic umbilic	2	4	$\frac{1}{2}x_1^2x_2 + \frac{1}{4}x_2^4 + \frac{1}{2}u_1x_1^2 + \frac{1}{2}u_2x_2^2 - u_3x_1 - u_4x_1$

Table 1.1 The seven elementary catastrophes

are constructed by replacing the function being unfolded by its negative. In effect, this means that the positions of maxima and minima are reversed as the control manifold is covered. The self duals are such because the negative sign can be produced by a change of coordinates. In effect, a review of Table 1.1 shows that functions which are wholly even-powered polynomials have duals which are different, while the rest, including at least one odd-powered term, are not, as replacing x by -x produces the required minus sign. (For example, in the case of the fold, x^3 simply becomes - x^3 on this transformation.) We illustrate briefly the concept of a dual by reference to the cusp. The potential function in Equation (1.12) is replaced by

$$z = -\frac{1}{4} x^4 + \frac{1}{2} u_1 x^2 + u_2 x \qquad (1.18)$$

and Figure 1.15 (literally turned upside down, as Amson points out) can be used to see what happens when maxima are turned into minima and vice versa. The only maxima for the cusp surface were on the middle sheet of the folded section, and so this part of the surface in the dual becomes the only set of minima and therefore the only observable states. Thus, there is a unique minimum inside the shaded area of the control manifold and no stable states outside it. The possible behaviours of the system are therefore less interesting than that of the basic cusp surface. This particular example of a catastrophe is also sometimes known as the false cusp.

High order catastrophes occur when families of functions with more than four parameters are involved. Some of these can now be classified and dealt with formally (see Poston and Stewart, 1978, for some examples). However, we will not proceed with these any further here, but merely note the potential excursions into greater complexity for the aspects of urban and regional systems which turn out to have this character.

Finally in this general review we note the existence and importance of what Poston and Stewart (1978) call *constraint catastrophes*. These arise as, in effect, extensions of Thom's

theorem. The theorem is concerned with maxima and minima
determined by points of the potential function where the derivative vanishes; if a model is constructed which includes
constraints, then observable minima may be determined by the
constraints rather than by vanishing derivatives. This point
is illustrated by the curve shown in Figure 1.16 which shows
the effect of a non-negativity constraint on a variable. In
this case, local maxima or minima often occur on the boundary
imposed by the constraint and the derivative of the potential
function does not vanish at that point. We will see later
that one of the fold examples concerned with retailing
structure arises in this way.

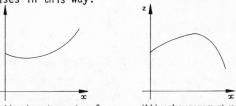

(a) Local maximum at $x = 0$ (b) Local minimum at $x = 0$

**Figure 1.16 Local maximum and local minimum created by the
constraint $x > 0$**

The catastrophes which can arise because of constraints
can be classified, after representing the constraints in some
canonical form, in a manner analogous to that used by Thom. A
preliminary treatment of this problem is offered by Poston and
Stewart (1978) and a more detailed account is to appear in Pitt
and Poston (1979).

CHAPTER 2

DIFFERENTIAL EQUATIONS AND BIFURCATION

2.1 Differential equations and non-gradient systems

Most dynamical systems of interest are not gradient
systems. That is, the corresponding differential equations
cannot be reduced to the form (1.2). Much is known about the
dynamics of such systems. Typically, for example, such
equations have a small number of isolated stable equilibrium
points and information about system behaviour is presented as
trajectories on state space (or phase) diagrams. The stable
points act as *attractors* (and correspondingly, unstable points
as *repellors*) and these points shape the contours of the trajectories in state space accordingly as we will see in examples
below.

In this chapter, we do not pursue or offer the standard
basic results about such systems, interesting and underdeveloped
though application of such methods are in urban and regional
studies. (These are available elsewhere - see for example,
Jordan and Smith, 1977.) Our main interest is to investigate
the nature and stability of equilibrium points and other solutions with respect to parameter changes. This is in the
spirit of catastrophe theory, but for non-gradient systems
where the theory does not apply directly. We will be particularly interested in the idea of *bifurcation* in more or less the
same sense as we have already used it: the existence of critical parameter values at which the nature of the solution to the
differential equation changes. However, there is no elementary and simple classification of possible 'cases' as in
catastrophe theory and so we proceed mainly by example rather
than by attempting to take the general discussion any further.

Chapter 2

We begin in Section 2.2 by reviewing the main concepts needed for the rest of the chapter.

2.2 Dynamical systems and solutions to differential equations : a sketch of basic concepts

2.2.1 A preliminary note on types of graphical presentation

One of the confusing aspects of the literature to the uninitiated is that information on the behaviour of dynamical systems is presented in graphical form for simple examples, but in a number of different ways. It is useful briefly therefore to alert the reader to some of the types of graph used, all of which will be deployed at some stage in our examples below.

(1) The most fundamental plot is perhaps that based on state space. For a system described by a vector of state variables, $\underline{x}$, this is a geometrical representation of possible trajectories in the space $(x_1, x_2, x_3, \ldots, x_N)$ in N dimensions say. One trajectory represents the behaviour of the system following given initial conditions. If there are only two variables, then this can be plotted in the usual Cartesian form as in Figure 2.1, where a typical trajectory is shown.

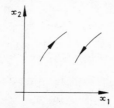

Figure 2.1 Trajectories in state space

(2) One of the basic diagrams used in Chapter 1 to give geometrical interpretations of catastrophe theory involved a plot in the higher dimensional space of the state variables and control variables (or parameters). In the case of a single state variable, x, and a single parameter, u, this could again be achieved in two dimensions as shown in Figure 2.2. As indicated, this may reveal the existence of a fold curve and the fold catastrophe which we met earlier. Such diagrams are usually restricted to plots of possible equilibrium points, and these are a very small subset of the points shown on

34

trajectories on state space diagrams. On the other hand, the
state-space plots are restricted to fixed parameter values.

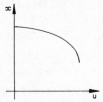

Figure 2.2   Trajectory in behaviour manifold in two dimensions

(3) The third kind of diagram involves plotting a state
variable, again say x, against some intermediate variable in an
attempt to illuminate a mechanism of change in the model: that
is, how the state variable changes against this intermediate
variable with perhaps additional conditions imposed which then
determine the position of the stable equilibrium points on such
a plot.

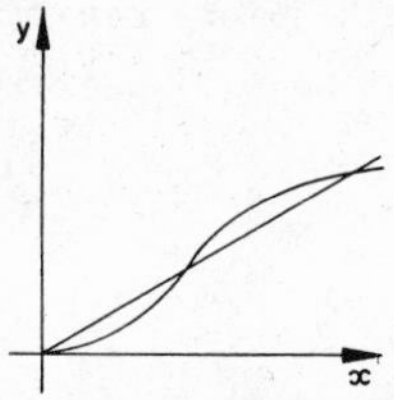

Figure 2.3   State variable plotted against intermediate variable

     An example of this is shown in Figure 2.3 where x is plotted
as a function of some variable y, and it is known that the
equilibrium points are the intersection of this curve and the
line x = uy, where u is some parameter.   It is then possible
to investigate the position of the equilibrium point as the
parameter u varies.   This, of course, is essentially the
example presented in Section 1.2.4 but it is worthwhile drawing
general attention to it as it turns up in a number of fields as
a mechanism for representing bifurcation behaviour.

(4) Another important plot to exhibit bifurcation behaviour is
of the value of a state variable, usually an equilibrium value,
against a parameter - say x against u as shown in Figure 2.4.

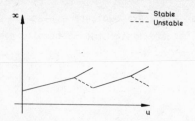

Figure 2.4 A state variable vs. a parameter , exhibiting bifurcation

The branching points represent bifurcation, and we discuss this in more detail in Section 2.2.5 below. This is a special case of (2) with one variable and one parameter singled out.

(5) Another basic diagram involves plotting a state variable against time. An example of oscillatory behaviour is shown in Figure 2.5. This emphasises the difference in form of plot for similar information in the different kinds of diagram. On a state-space plot, such oscillatory behaviour would be represented by a closed orbit, as we will see with one of the examples below.

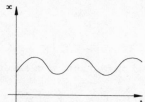

Figure 2.5 A state variable plotted against time

(6) A more unusual plot is of the time derivative of a state variable against the state variable itself, say of $\dot{x}$ against x. An example is shown as Figure 2.6 (following Mees, 1975). The

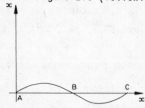

Figure 2.6 A state variable plotted against its time derivative

36

intersections of the curve with the horizontal axis show the positions of the equilibrium points and also give a quick indication of their stability. In the figure, A and C are unstable, while B is stable.

The problem to be borne in mind in all these cases is that, typically, plots are focusing on one or two variables when the system is represented in a higher dimensional space. These diagrams are then slices of corresponding higher dimensional ones. They can still offer useful insights, but this underlying complicating feature should be borne in mind.

2.2.2 Equilibrium points and trajectory sketching

Differential equations can always be written in first order form by the addition of new variables. For example, if $\ddot{x}$ appears in an equation, a new variable y can be introduced, set equal to $\dot{x}$; then $\dot{y}$ appears in the equation instead of $\ddot{x}$. We can assume therefore that in general, a set of differential equations for state variables $x_1, \ldots, x_n$ (and parameters $u_1, \ldots, u_m$) can be written in the form

$$\begin{aligned} \dot{x}_1 &= f_1(x_1, \ldots, x_n, u_1, \ldots, u_m) \\ \dot{x}_n &= f_n(x_1, \ldots, x_n, u_1, \ldots, u_m) \end{aligned} \tag{2.1}$$

and this can be written in vector form more conveniently and using an obvious notation as

$$\dot{\underline{x}} = \underline{f}(\underline{x}, \underline{u}) \tag{2.2}$$

(We should emphasise that, in practice, it may be difficult to get equations into this reduced and canonical form.)

The equilibrium points of such a system occur when $\dot{\underline{x}} = 0$, that is, when

$$\underline{f}(\underline{x}, \underline{u}) = 0 \tag{2.3}$$

So, in order to find these points, the simultaneous equations (2.3) have to be solved. These will usually be non-linear, and this may be a very difficult exercise. If it can be done analytically, then it is always useful; if not, then the equilibrium points may be found by numerical simulation on a computer. This usually involves an iterative procedure which, in effect, is a 'sketch' of the trajctories which lead to equilibrium points. Sometimes it is feasible to get insights by hand computations of trajectories. An alternative and instructive method is to use geometrical sketches for a state space diagram of the individual curve (2.4) (in two dimensions) - or surfaces (in higher dimensions) - so that the intersection of the curves are the possible equilibrium points. We will see examples of this method below, beginning with examples of typical trajectories in Section 2.2.4.

The stability of equilibrium points, or the direction of progress along trajectories can be investigated by looking at the sign of $\dot{x}$ in a neighbourhood of such points (*cf*. (5) in Section 2.2.1 above), or by examining Hessian matrices - see, for example, Jordan and Smith (1977).

2.2.3 Static models embedded in dynamic frameworks

This is a useful place to note the important point that many models, especially in relatively underdeveloped fields like urban and regional studies, are *stated in terms of equilibrium points only*. That is, they assume equilibrium, and the equations of the model represent these points directly without any reference to rates of change or to how the equilibrium is achieved. Bellman (1968) noted the existence of this problem in a rather general way and introduced the notion of *embedding* such static models in a dynamic framework. In effect, this involves arranging the equations of the static model in the form of Equation (2.3) - which may not be a trivial matter - and then setting up the differential equations in the form (2.1). A number of examples of this type of development will be presented later.

One example of this is provided by urban retailing structures as outlined in Section 1.2.4 above. The equilibrium conditions for $\{W_j\}$ were given by Equation (1.6). They can be written as

$$D_j - kW_j = 0 \qquad\qquad (2.4)$$

which are now of the form (2.3) for the state variables $\{W_j\}$.
Differential equations of the form (2.1) are thus

$$\dot{W_j} = \varepsilon(D_j - kW_j) \qquad\qquad (2.5)$$

for a suitable constant, ε. Since D_j is a complicated function of $\{W_k\}$, these are complicated non-linear equations in
variables, $\{W_j\}$ and $\{\dot{W_j}\}$ and these are pursued in depth in
Chapter 5 below.

2.2.4 Basic types of trajectory

It is found that, typically, there is a relatively small
number of different types of trajectory. As noted earlier, it
is most common for a system to have a small number of single
equilibrium points. If they are stable, then trajectories
lead into them; they are, as we noted earlier, *attractors*. If
they are unstable, then trajectories are repelled by such
points; they are *repellors*. When there are two or more state
variables, the equilibrium points may be *saddle points*, which
represent a special kind of instability; most trajectories are
repelled by such points, but there can be two trajectories
which pass through the saddle and these play an important role
in sketching trajectories in general. They separate the state
space into two regions with trajectories on each side being
directed to different stable equilibrium points. For this
reason, such a trajectory is known as a *separatrix* and it
plays an important role in bifurcation behaviour.

To fix ideas, consider a system described by state
variables x and y. First, we distinguish two kinds of
behaviour in the neighbourhood of a stable equilibrium point.
These are shown on a state-space plot in Figure 2.7. In
case (a), the trajectories lead directly to the equilibrium
point and represent an exponential convergence; in case (b),
the trajectories spiral into it and represent oscillating convergence. Typical time plots for x against t are shown for
the two cases in Figure 2.8. There are corresponding plots

39

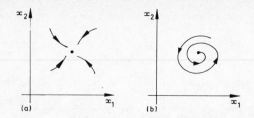

Figure 2.7 Stable equilibrium points in state space

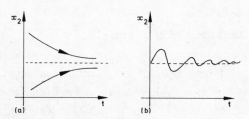

Figure 2.8 Time plots representing progressions to a stable equilibrium point

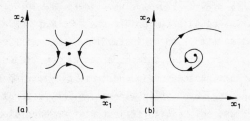

Figure 2.9 Unstable equilibrium points in state space

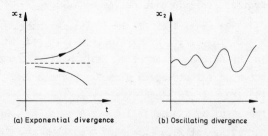

Figure 2.10 Time plots representing divergence from an unstable equilibrium point

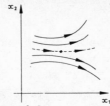

Figure 2.11 A saddle point

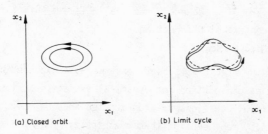

(a) Closed orbit   (b) Limit cycle

Figure 2.12 Periodic trajectories in state space

for unstable equilibrium points in Figures 2.9 and 2.10.
Saddle points behave like unstable case (a) points, but with
the addition of the trajectories which form the separatrix as
shown in Figure 2.11.

   If behaviour is neither convergent to a stable point, nor
divergent, then it may be *periodic*.   There are two basic types
as shown in Figure 2.12.   Case (a) is *closed orbit* periodicity,
when the trajectory never leaves one of many possible such
orbits (the particular one being determined by the initial conditions).   Case (b) is *limit cycle* behaviour: a typical
trajectory winds in and out of a closed orbit, and may become
asymptotically close to it.   It turns out that closed-orbit
behaviour is structurally unstable while limit-cycle behaviour
is structurally stable.   This notion is elaborated to
illustrate bifurcation in Section 2.2.5 below.

   Finally, there are examples of system behaviour characterised by neither equilibrium points, stable or unstable, nor
by oscillating behaviour of any regular periodicity.   Such
behaviour is called *chaotic* and is demonstrated by irregular

looking time plots of state variables. We will see an example in Chapter 8.

A particular (complicated) system may exhibit a number of different kinds of solution for different starting values of the variables and for different parameter values. So a state-space diagram may be a mixture of trajectories related to different kinds of equilibrium values, and may change character as the parameters change. We shall discuss this further below in the context of bifurcation.

2.2.5 Bifurcation

In the preceding subsections, we have explored how the main types of solution for dynamical systems described by differential equations are constructed and how to get some insights by representing them graphically. It will already be clear by implication that the possible types of bifurcation behaviour are richer than in catastrophe theory. We now summarise the main types in a broad way and then illustrate them with examples in the rest of the chapter.

First, we note that the solutions (for equilibrium points) to Equations (2.3) will, typically, involve multiple solutions because of any non-linearities in the functions $\underline{f}$. Hence, the manifold of equilibrium solutions in the space of $(\underline{x}, \underline{u})$ variables (where $\underline{u}$ are the parameters as usual) will be folded. This can lead to the same broad kinds of bifurcation as in catastrophe theory.

Secondly, we observe that the types of solution to the differential equations can be determined by parameter values. There can be critical parameter values at which a stable equilibrium point becomes unstable or disappears; or at which a periodic solution could disappear and be replaced by a stable equilibrium point - or vice versa (which is known as the Hopf bifurcation). In theory, all the possible interchanges between the kinds of solution (or trajectory) listed in Section 2.2.4 are possible and it is useful to be alert to this in applied work.

Finally, we also noted a completely different type of possible bifurcation. If a system is disturbed from an equilibrium position and moves to a non-equilibrium state in the neighbourhood of a separatrix in state space, then if the separatrix is crossed, the return to equilibrium could be to a state other than the original one.

2.3 Examples

2.3.1 Introduction

In this section, we present a range of examples which illustrate the main concepts of bifurcation outlined above. The examples are well-known equations, often from the ecological literature where the first applications were developed. First, in subsection 2.3.2, we examine a range of single-variable growth equations and then the famous Lotka-Volterra prey-predator equations which illustrate a two-variable interacting system. Another variant of these equations is examined in subsection 2.3.4. It is possible in this context to show how bifurcation properties can be deduced from a formulation involving general functions and which includes separatrix crossing. Finally, in subsection 2.3.5, we examine simultaneous logistic growth equations in N variables. The study of these builds on the earlier simpler examples and provides the basis for the main urban applications in Chapter 5. Most of the framework of the argument follows Maynard Smith (1974) and Hirsch and Smale (1974).

2.3.2 Growth equations

The simplest example of bifurcation arises from the differential equation which describes exponential growth. This argument is taken from Hirsch and Smale (1974). The equation is, of course,

$$\dot{x} = ux \tag{2.6}$$

The behaviour of this system is qualitatively different for positive and negative u as shown (with the addition of the u = 0 case) in Figure 2.13. For u > 0, there is exponential growth; for u < 0, there is exponential decline asymptotic to

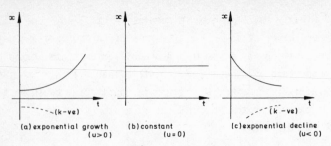

Figure 2.13 Plots of x= Ke^{ut} for varying u

zero and for u = 0, the function is of course constant. In the
first two of these cases, there is also the negative exponential growth or decline corresponding to a negative value of the
constant of integration in the solution

$$x = Ke^{ut} \qquad (2.7)$$

The behaviour in these various cases is visibly qualitatively
different, and it is clear that u = 0 is the critical point at
which bifurcation of behaviour takes place.

  Few systems can sustain indefinite exponential growth.
Often, growth is exponential in early stages but is then limited
by an upper bound. This is a common feature of ecological and
urban systems where the bound is created by resource limits.
The differential equation which represents this process is
achieved by modifying the growth coefficient u in Equation (2.6)
so that it declines as the population grows.   This can be done
by introducing an additional factor (1 - x/D):

$$\dot{x} = u(1 - x/D)x \qquad (2.8)$$

and this has the solution plot shown in Figure 2.14. It can
easily be seen that D is the upper bound of growth (because $\dot{x}$
is zero at x = D and negative for x > D) and so it is sometimes
known as the carrying capacity of the system. Indeed, x = D is
the only stable equilibrium point, since although $\dot{x}$ = 0 when
$\dot{x}$ > 0 for small positive x.

  By rescaling the parameter u, Equation (2.8) can be written
in the more convenient form

44

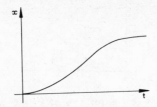

Figure 2.14 Logistic growth

$$\dot{x} = u(D - x)x \qquad (2.9)$$

This basic model can be further extended by making D a function of x - in effect introducing feedback between x and the carrying capacity. The equation can then be written

$$\dot{x} = u[D(x,\lambda) - x]x \qquad (2.10)$$

where λ is an example of a new parameter. If $D(x)$ is a non-linear function, then there will be new equilibrium points which are the solution of

$$D(x,\lambda) = x \qquad (2.11)$$

and these may appear or disappear at critical values of λ, the bifurcation points. The twin properties of feedback and non-linearity in this kind of model are usually generators of interesting bifurcation properties.

It is also useful to explore other forms of growth equation. Consider, for example, the possibility of replacing the factor x in Equation (2.8) by a factor x^n and seeing what happens for different values of n:

$$\dot{x} = u(1 - x/D)x^n \qquad (2.12)$$

An important case which turns up frequently is n = 0. In this case, x is finite and non-zero at the origin and so the pick-up there is faster than for the logistic curve. For n < 0, the gradient of pick-up at the origin becomes infinite while for n > 0, the gradient is always zero as in the logistic case; for n > 1, the inflexion is more accentuated. A variety of plots of x against t are sketched in Figure 2.15.

45

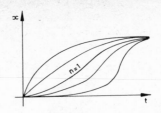

Figure 2.15 Plots of the solutions of $\dot{x} = \varepsilon (D-x)x^n$

One particular feature of Equation (2.12) should be noted at this stage: if the variable x is zero at any time (for example, initially) then it remains zero.   This is obviously disabling if we are trying to model systems evolution, but it can be overcome using the theory of fluctuations which is taken up later.

2.3.3 Competition 1 : The Lotka-Volterra equations

The next step in the argument is to consider the growth of more than one population where there is some interaction between them.   The most famous of these simultaneous equations was originally developed indpendently by Lotka and Volterra in the 1920s and has been modified in a great variety of ways since then.   They represent the interaction between prey and predator. We consider them here in the form discussed by Maynard Smith (1974);   there is a slightly different discussion offered by Hirsch and Smale (1974).   Let x_1 be the population of a prey species and x_2 that of a predator species.   Then, the x_1 population is supposed to grow logistically but with an additional decline term due to being eaten by the x_2 population.   Its equation is therefore

$$\dot{x}_1 = (a - bx_1 - cx_2)x_1 \qquad (2.13)$$

The predator population is assumed to grow in proportion to the volume of x_1 which is available and otherwise to decline at some known rate:

$$\ddot{x}_2 = (- e + fx_1)x_2 \qquad (2.14)$$

a, b, c, e and f are suitable parameters.

We can use this example to illustrate how to explore the nature of equilibrium points and trajectories on phase space diagrams. The equilibrium points are the solutions of

$$a - bx_1 - cx_2 = 0 \tag{2.15}$$

and

$$x_1 = 0 \tag{2.16}$$

which intersects with those of

$$- e + fx_1 = 0 \tag{2.17}$$

and

$$x_2 = 0 \tag{2.18}$$

We concentrate on the interesting equilibrium point where both solutions take a non-zero value and so ignore the axes (2.16) and (2.18). The lines (2.15) and (2.17) are shown plotted on Figure 2.16. The lines divide the positive quadrant of the

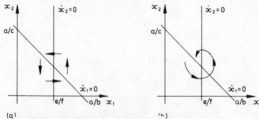

(a)   (b)

Figure 2.16 Stable equilibrium points for the Lotka –Volterra equations

state space diagram into four quadrants and the next step is to add arrows on each quadrant which show the direction of change of the variables. (This can be carried out in practice simply by calculating $\dot{x}_1$ and $\dot{x}_2$ at a number of sample points: $\dot{x}_2/\dot{x}_1$ then gives the gradient of the trajectory at each such point.) This is done in the figure, using information from Equation (2.13) and (2.14). These indicate the nature of a typical trajectory which is shown on part (b) of the figure: it is like that in Figure 2.1, case (b), and 'spirals in' to the equilibrium point, which is thus seen to be stable.

Various interesting points can be made about the nature of this stability. First, if the slope of the line $\dot{x}_1 = 0$ changes to the sort of position shown on Figure 2.17, then the point

47

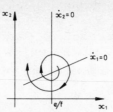

**Figure 2.17 An unstable configuration for
the Lotka–Volterra equations**

would be unstable and the trajectory now spirals outwards, as
shown. This could only happen, of course, if the coefficient b
in Equation (2.13) was negative, which would imply an accele-
rated exponential growth - the growth rate being proprotional
to itself. It is hardly surprising that such a system would
exhibit spiralling divergence. However, it enables us to make
an interesting mathematical point.

Secondly, we note that the two lines shown in Figure 2.16
need not intersect as shown in the positive quadrant. The condition for this can be seen from Figure 2.16 to be

$$a/b \; > \; e/f \tag{2.19}$$

If this does not hold, we get the situation shown in
Figure 2.18. (Analysis, as indicated by arrows on the figure,
shows that the point on the x_1 axis is now stable.) We can

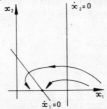

**Figure 2.18 An alternative Lotka –Volterra configuration
with the predator eliminated**

present this information in another and more familiar way by
supposing that all the parameters except one in (2.19) are constant. Let the single variable parameter be b and define a
convenient control parameter as $u_1 = 1/b$. It can then be seen
from the figure that x_2 as a function of u_1 can be plotted as
in Figure 2.19, and that this represents a fold catastrophe
relationship between x_2 and u_1.

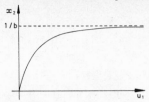

Figure 2.19 A fold catastrophe for parameter change in the Lotka-Volterra system

The third feature to note is that there is an interesting critical point of b - in fact zero. This is interpreted as meaning that there is no limit imposed on the prey population growth by itself. In this case, the trajectories in phase space become periodic as was shown in Figure 2.12(a). This case shows the nature of structural *in*stability: given Equations (2.13) and (2.14) with b = 0 if a term $-bx_1^2$ is suddenly introduced with b small but non-zero, then the state space diagram reverts to Figure 2.16 form.   And this also illustrates bifurcation between solutions based on stable equilibrium points and a periodic solution.

2.3.4 Competition 2 : fixed resources

Consider again a system of two populations, x_1 and x_2, but which are now competing for fixed resources - say different species competing for a fixed food supply - rather than in a prey-predator relationship.   We will see later that there are many more possible interpretations and applications - such as retailers competing for consumers in a market, transport mode operators for a fixed supply of passengers and so on. To see the simplest way of studying the dynamics of such a system, we again follow Maynard Smith (1974). Each population is supposed to grow according to a kind of logistic equation of the form of (2.13) in the prey-predator model:

$$\dot{x}_1 = (a - bx_1 - cx_2)x_1 \qquad (2.20)$$

$$\dot{x}_2 = (e - fx_1 - gx_2)x_2 \qquad (2.21)$$

In effect, each population's growth is limited by a term reflecting its own use of resources and the effect of growth of

49

Chapter 2

the other population which is competing for the same resources. Again, we concentrate on equilibrium points with both x_1 and x_2 non-zero by plotting the line which make the right hand side of Equations (2.20) and (2.21) zero. These are given for four cases in Figure 2.20. An examination of directions of trajectories in the neighbourhood of the equilibrium points shows

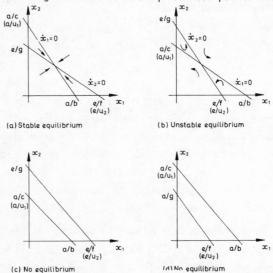

Figure 2.20 **Alternative state-space configurations for the competition-for-resources (C-F-R) model**

that in the case (a) the equilibrium is stable and in case (b), it is unstable. In cases (c) and (d), there is no equilibrium point in the positive quadrant. In this instance, the four cases can be distinguished by the variation of two control parameters, say u_1 and u_2 which we take to be c and f respectively, thus measuring the relative strengths of the competition. (We assume that the other parameters remain constant.) If we define new constants A_1 and A_2 such that

$$A_1 = ag/e \qquad (2.22)$$

$$A_2 = be/a \qquad (2.23)$$

then the inequalities which distinguish between the four cases are

(a) $u_1 < A_1$ and $u_2 < A_2$ (2.24)
(b) $u_1 > A_1$ and $u_2 > A_2$ (2.25)
(c) $u_1 > A_1$ and $u_2 < A_2$ (2.26)
(d) $u_1 < A_1$ and $u_2 > A_2$ (2.27)

These regions of parameter space are shown on Figure 2.21, which
is a plot of the control manifold for this system showing the
stable and unstable and no-equilibrium cases around a critical
point given by $u_1 = A_1$ and $u_2 = A_2$. It is interesting to see
how, for this non-gradient system, such a plot does not take any
of the shapes familiar in catastrophe theory, it is not an
obvious transformation of the cusp projection of the control
manifold of the cusp catastrophe (which has two control
parameters), for example.

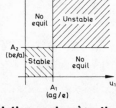

**Figure 2.21 Different ‘solution regions’ on the control
manifold for the C-F-R model**

There is a more general treatment of the 'competition for
resources' problem presented by Hirsch and Smale (1974). This
illustrates different bifurcation behaviour, and also that
reductions can be made about the dynamics of a system whose
properties have been specified only in terms of functions with
broadly-specified attributes. The equations are taken in the
form

$$\dot{x}_1 = M_1(x_1, x_2, \underline{u})x_1 \qquad (2.28)$$

$$\dot{x}_2 = M_2(x_1, x_2, \underline{u})x_2 \qquad (2.29)$$

where $\underline{u}$ is a vector of parameters. We will see in a detailed
example later how these results can be extended for an
n-dimensional state vector (though with the price of not being
able to visualise phase diagrams). The nature of the competition is then specified through a set of assumptions about the
functions M_i:

51

Chapter 2

(1) If either population increases, the other declines

$$\frac{\partial M_1}{\partial x_2} < 0 ; \qquad \frac{\partial M_2}{\partial x_1} < 0 \qquad (2.30)$$

(2) There is a limit to the growth of either population: that is there exists a constant K such that

$$M_1 < 0 \text{ and } M_2 < 0 \qquad (2.31)$$
$$\text{if } x_1 > K \text{ or } x_2 > K$$

(3) If one species is absent, the other grows to a certain point but not beyond it:

$$M_1(x_1,0) > 0 \text{ for } x_1 < a_1, \qquad (2.32)$$
$$M_1 < 0 \text{ for } x_1 > a_1$$

$$M_2(0,x_2) > 0 \text{ for } x_2 < a_2, \qquad (2.33)$$
$$M_2 < 0 \text{ for } x_2 > a_2$$

As usual, we can proceed to look for equilibrium points on the basis of solving the equations:

$$M_i(x_1,x_2,\underline{u}) = 0 \qquad (2.34)$$

Let the set of (x_1,x_2) for which M_1 vanishes form the curve A_1 and the set for which M_2 vanishes be A_2. These curves are shown on Figure 2.22. M_i is greater than zero inside A_i and less than zero outside.

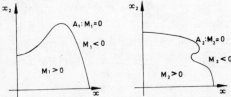

Figure 2.22 C-F-R model: plots of $\dot{x}_1 = 0$ (A_1), $\dot{x}_2 = 0$ (A_1)

If the curves do not intersect, then they can take the form shown in Figure 2.23 to which a number of trajectories have been added. The equilibrium points are $(0,0)$, $(a_1,0)$ and $(0,a_2)$ and in this particular configuration, most trajectories tend to $(0,a_2)$.

52

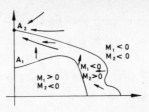

Figure 2.23 C-F-R model: A_1, A_2 non-intersecting

If the curves do intersect, then the assumptions can be used to find the signs of $\dot{x}_1$ and $\dot{x}_2$ in each region of the phase diagram. This information is incorporated into Figure 2.24. In this case, the stable points are B and P as shown and Q is a saddle point. The two trajectories which pass through Q are shown on the diagram, and as noted earlier they form separatrices: all trajectories above them go to B; all those below to P. This provides a specific illustration of the corresponding kind of bifurcation: if the system is disturbed to a point

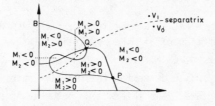

Figure 2.24 C-F-R model: A_1, A_2 intersecting

near the separatrix - say to V_0 or V_1 as shown, then this may lead to a system jump. For example, if it goes from P to V_0, then it will return to P; but if for a slight additional change, it goes to V_1, then it will return to B and this will be seen as a jump.

The more conventional kind of bifurcation involves the appearance and disappearance of equilibrium points. In this example, the difference between Figures 2.23 and 2.24 is created by differences in parameter values, $\underline{u}$. There will be one or more critical values of $\underline{u}$ which give the transition between one case and another. An explicit example of this will be seen in Chapter 5 when the functions M_i are made more specific.

The possibility of these kinds of bifurcation can be seen and established on the basis of only a very general knowledge of the differential equations involved.

2.3.5 Logistic growth equations for interacting populations

We can now combine our discussion of logistic growth equations and models of interacting populations. In Section 2.4.2, we showed how logistic growth could be represented for a single population and the most general form of equation was presented as (2.10). An obvious extension for an n-species population is:

$$
\begin{aligned}
\dot{x}_1 &= \varepsilon_1 \left[D_1(x_1, x_2, \ldots, x_n, u_1, u_2, \ldots, u_m) - x_1 \right] x_1 \\
\dot{x}_2 &= \varepsilon_2 \left[D_2(x_1, x_2, \ldots, x_n, u_1, u_2, \ldots, u_m) - x_2 \right] x_2 \\
&\vdots \\
\dot{x}_n &= \varepsilon_n \left[D_n(x_1, x_2, \ldots, x_n, u_1, u_2, \ldots, u_m) - x_n \right] x_n
\end{aligned}
\tag{2.35}
$$

where there is a carrying capacity, D_i, for each population i shown as a function of all the populations, $x_1, x_2, \ldots, x_n$, and of m parameters $u_1, u_2, \ldots, u_m$. The set of parameters $\varepsilon_1, \varepsilon_2, \ldots, \varepsilon_n$ which determine the relative rates of growth completes the picture. The interaction between the populations is represented by the forms of the function D_i. The equilibrium points are the solutions of

$$
D_i(\underline{x}, \underline{u}) = x_i, \quad i = 1, 2, \ldots, n
\tag{2.36}
$$

and when the functions D_i are non-linear, this can obviously generate interesting bifurcation behaviour. Equations (2.35) and (2.36) can easily be modified to incorporate alternative growth paths by the replacement of the x_i factor by x_i^n as in Equation (2.12).

The prey-predator equations and the 'competition for 'resources' equations can now be seen as special cases of (2.35). For the former, taken n = 2 and

$$D_1(x_1,x_2,a,b,c,e,f) = (a - cx_2)/b \qquad (2.37)$$

and

$$D_2(x_1,x_2,a,b,c,e,f) = x_2 - e + fx_1 \qquad (2.38)$$

with

$$\varepsilon_1 = b \qquad (2.39)$$

and

$$\varepsilon_2 = 1 \qquad (2.40)$$

For the latter, take n = 2 and

$$D_1(x_1,x_2,a,b,c,d,e,f) = (a - cx_2)/b \qquad (2.41)$$

$$D_2(x_1,x_2,a,b,c,d,e,f) = (e - fx_1)/g \qquad (2.42)$$

with

$$\varepsilon_1 = b \qquad (2.43)$$

$$\varepsilon_2 = g \qquad (2.44)$$

2.3.6 Further extensions: linked subsystems and fluctuations

Suppose one set of populations is given by the vector $\underline{x}$ and has dynamics described by Equations (2.35). Let $\underline{y}$ be a p-vector representing a new set of populations. Then Allen *et al.* (1978), whose work we will return to in Chapter 6, introduced a set of equations of the form

$$\dot{y}_i = \rho_i[E_i(y_1,y_2, \ldots, y_p,v_1,v_2, \ldots, v_q) + \sum_k R_k x_k - y_i]y_i,$$
$$i = 1, \ldots, p \qquad (2.45)$$

for suitable parameter sets $\underline{\rho}$, $\underline{v}$ and $\underline{R}$. The holding capacities E_i are being added to by functions of the other populations through the term $\sum_k R_k x_k$. There is thus a representations of feedback between the two major subsystems.

A second feature of the work of Allen *et al.* (1978) is the explicit introduction of stochastic terms - fluctuations. This is a useful step in two senses. First, random perturbations are an important feature of real systems. Secondly, they play an important role mathematically in bifurcation theory. For models represented by equations of the form (2.34), possibly with (2.45) added, if any variable x_i (or y_i) is zero initially, then it remains zero because of the factor x_i. Allen *et al.* introduce positive values of such variables as small fluctuations, introduced using a random number generator, and this creates the possibility of growth. More generally, such fluctuations can be introduced away from equilibrium during the simulation of a system's growth. In effect, when the system is next to a stable equilibrium point, they will be ignored, but in the neighbourhood of a critical point, the random fluctuations determine which branch the system takes.

CHAPTER 3

APPLICATIONS OF DYNAMICAL SYSTEMS THEORY: A SURVEY OF APPROACHES

3.1 Introduction

In the rest of this book, a wide range of applications of the methods of dynamical systems theory is presented which covers both urban studies and a number of other disciplines. These are based on the ideas outlined in the first two chapters. It is useful at this stage to take stock before embarking on the main chapters and to review the nature of the types of application which are possible.

First, in Section 3.2, we examine the relationship between catastrophe theory and the theory of differential equations and begin to see that, in applied work, the two branches will work hand in hand. In Section 3.3, we follow through an idea introduced earlier that it is important to study explicitly the different rates of change of variables in a model and to use this as the basis for distinguishing different types of variables. This idea is used as a basis to draw together some notions on system description.

Zeeman (1977A) introduced a very useful 'six levels of approach' to applications of catastrophe theory and differential equations and these are presented in Section 3.4. Two issues are then considered briefly in the following section. First, the mathematics of catastrophe theory is often said to be qualitative rather than quantitative and we explore the meaning behind such comments. Secondly, we investigate whether the model building style to be adopted is inductive or deductive. In Section 3.6 we note the creation of some new styles of planning application generated by the study of bifurcation.

Chapter 3

3.2 Differential equations and catastrophe theory

Catastrophe theory as such was presented in Chapter 1 as being concerned largely with the topology of surfaces of possible equilibrium states of gradient systems. The behaviour of systems is represented by trajectories on such surfaces. The path taken by a particular trajectory is determined by the way in which the parameters, $\underline{u}$, vary (and the particular trajectory by the initial conditions of all the variables - $\underline{x}$ and $\underline{u}$). We saw that the differential equations governing the $\underline{x}$-variables took the form

$$\dot{\underline{x}} = - \text{grad } f(\underline{x},\underline{u}) \qquad (3.1)$$

and we now note that the particular path on the equilibrium surface can be given by specifying differential equations for the u-variables, say as

$$\dot{\underline{u}} = \underline{g}(\underline{u},\underline{y}) \qquad (3.2)$$

for some suitable vector of functions, $\underline{g}$, and where the y's are a set of constants which have been introduced. Note that the u-variables' equations will not necessarily be gradient equations.

When we describe the behaviour of a system using catastrophe theory, we nearly always assume that the system stays on the equilibrium surface. This means that the speed of return to equilibrium of equations (3.1) is considered to be very fast compared to the corresponding situation for the parameter equation (3.2). Hence the former equations are known as the *fast equations* and the latter as the *slow equations*. The surface of equilibrium variables is known as the slow manifold. If there is any disturbance from equilibrium, the system is assumed to move rapidly back to equilibrium via what Zeeman (1977A) calls the 'fast foliation'. These are trajectories which are 'off' the manifold but which lead directly on to it at the same $\underline{u}$-value which held before the perturbation. If the $\underline{u}$-equations (3.2) have an equilibrium point, then the system moves more slowly on the manifold until this is reached[1].

3.3 Relative speeds of change : variables , parameters and constants ; system description

The first step in applied work is obviously the definition
of variables to describe the system of interest. The preceding
sections help us to identify an important element in this: the
division of variables according to speeds of change.
Essentially, the state variables, that is, the x-variables, are
'fast' while the control variables (or parameters), the
u-variables, are 'slow'. We also introduced the notion of
'constants', y. There is bound to be some element of arbitrari-
ness in the way these distinctions are made, and indeed this
often forms an interesting research question. It is being
argued here that the typical time rates of change of variables
provide the most important criteria, but the alternative names
we use for the different kinds of variable also suggest other
implicit criteria. For example, 'control variables' may be
literally such and their role in this respect may conflict with
'rate-of-change' definitions. We also noted from catastrophe
theory that the theory can handle large numbers of state
variables but only up to four control variables. It is therefore tempting to classify variables so that these conditions
are met. Indeed, sometimes, we may want to go further:
although the theory can handle any number of state variables,
in practice, the transformations involved in getting these into
canonical form are very difficult and so it is often more convenient to attempt to restrict the number of state variables as
well as the number of parameters.

One element of the preceding discussion points to a
solution to some of these difficulties - one which is not yet
much evident in applied work. When the differential equations
for the u-variables were introduced, it seemed intuitively
correct to make those equations functions of 'parameters' which
we treat as constants for this particular problem. What this
means in practice is that these parameters have rates of change
which are even slower than those of the u-variables. It would
easily be possible to extend this argument and to develop whole
hierarchies of variable classifications.

Chapter 3

Two ideas follow from this. At the practical level, it offers a way of 'slicing up' the problem: by putting variables into the y-category, we make the primary (x,u)-problem simpler, and this may be a way of handling complexity. At the theoretical level, it offers the possibility of new mechanisms of change. If the u-variables first, and then the y-variables were in turn described by differential equations representing gradient systems, then the whole system would be modelled by nests of catastrophe mechanisms. This kind of hierarchy could be extended almost indefinitely if necessary. If the u- (or y-) variables were not gradient systems, then the other bifurcation results of Chapter 2 would apply and so again we would have the whole system described by nested bifurcation mechanisms.

In summary, then, we are arguing that the most important principle of system description is the classification of variables by 'rate-of-change' into state variables, parameters and constants, for example, or possibly a more extensive hierarchy. These notions may be modified by the role of some variables as control variables and also by practical considerations. The traditional considerations of system description will also continue to apply. For example, it will be important to identify useful *scales* correctly as we will see in the following chapters which are organised in this way. Other conventional considerations are reviewed more fully elsewhere - for example in Wilson (1974, Chapter 11).

3.4 Levels of approach

We noted earlier that Zeeman (1977A) introduces the idea of six levels of approach to catastrophe theory (and indeed to dynamical systems theory more generally). They can be taken, in total, to represent what is involved in modelling the behaviour of a complicated system. They also represent steps which are themselves of an increasing order of technical difficulty and we will often have to be satisfied by the insights or approximate models which can be constructed using

the early stages only.   We first summarise the levels and then
discuss them in turn.   They are:

(1) The study of the equilibrium surface and its singularities.

(2) The specification of the fast dynamic - the differential
     equations for the $\underline{x}$-variables.

(3) The specification of the slow dynamic - the differential
     equations for the $\underline{u}$-variables.

(4) The representation of any feedback between the fast and
     the slow variables.

(5) The recognition of noise - which can cause fluctuations.

(6) The modelling of diffusion processes - which, in
     catastrophe theory, often involves taking the control
     variables as space-time variables.

   In the light of the discussion of the preceding section it
might also be worthwhile adding another step between (1) and (2)
and then possibly amending steps (2) and (3) in the light of the
outcome of this.   It could be presented as:

(1A) The identification of sets of variables with different
     relative speeds of change (which would be fast and slow if
     there were only two) and the associated nesting of the
     coresponding mechanisms of change.

   However, we continue the discussion in terms of Zeeman's
categories.

   The first step has been dealt with in general terms in
Chapter 1. We found there that equilibrium state manifolds may
not always be smooth and this leads to the possibility of
unusual system behaviour such as jumps, divergence or hysteresis.
In much urban and regional modelling to date, there has been an
emphasis on the calculation of equilibrium states and catastrophe theory does offer new insights.   There is a sense in
which 'comparative static' approaches can be made more interesting again and take their place within dynamical theory - albeit
as a first level, or step, as implied here.   There has also
been an emphasis in much applied urban work on the role of constraints, and such ideas too, as we saw briefly in Chapter 1,
can be incorporated within catastrophe theory.

61

Chapter 3

The second and third steps are each concerned with the specification of differential equations. Sometimes, this involves embedding a comparative static model, in the manner of Bellman (1968), in a dynamical framework made up of such equations. It is always important to calculate the equilibrium points, however, because they often shape the trajectories of the system in state space.

Bifurcation properties arise for at least three reasons: first, when the equilibrium surface is folded; secondly, when the nature of the solutions to one or other (or both) sets of differential equations changes at critical parameter values; and, thirdly, due to separatrix crossing after a disturbance. Thus, the general shape of the model - or at least the main possibilities of interesting behaviour - are determined when even the first three steps of the model building process are used.

The fourth level involves adding a further degree of complexity to the differential equations: the incorporation of feedback between the two main sets of equations. Typically, this will involve making the slow equations functions of some x-variables as the interaction in the other direction is already present. Since this usually introduces new non-linearities, it also introduces new bifurcation possibilities.

The remaining steps are concerned with adding realistic complexity to the models. The addition of fluctuations turns out to have interesting theoretical consequences which we will explore later in the context of the work of Prigogine and his colleagues in Brussels. It reflects an important element of reality of course. The sixth level, the representation of diffusion processes by taking control variables as space and time, has not been used in the urban and regional modelling to date. But since progress has been made with the other steps, and since it is an idea which lends itself naturally and directly to geographical theory, this is a step forward we can expect relatively soon.

62

3.5 Qualitative vs. quantitative; inductive vs. deductive

It is often argued, particularly by many of the mathematicians already cited, whose main work lies in catastrophe theory but also in relation to differential equations and non-gradient systems, that the mathematics is more important for its qualitative insights than for the more traditional style of detailed quantitative modelling. Indeed, most of the applications to date have been essentially qualitative, though this is now beginning to change.

It is in relation to this issue that the greatest controversies rage about catastrophe theory. Does the theory say *enough* to justify the conclusions which are drawn? The essence of a qualitative argument is something like the following. Variables are defined to describe some system of interest including an approriate number of control variables. This number can then be used to determine, from Thom's theorem, the nature of the worst kind of degenerate singularity which can occur. The canonical form of surface, or usually at least something closely related to it, is then used to draw a number of conclusions about the possible, and often asserted as likely behaviour of the system. Critics argue, firstly, that the real systems involved should often be represented by equilibrium surfaces which are complicated transformations 'away from' the canonical forms which are used; and secondly, that in any case Thom's theorem only applies locally and this in itself is enough to cast doubt on many of the applications.

In this book, we do not enter this argument in any detail. It has to be fought out case by case for each application. In the next chapter, however, a number of examples of this generalised qualitative argument from the geographical literature are included. The principle adopted here is that the representation of detailed mechanisms of change should be sought and this usually means building a quantitative model in the traditional sense. The main use of catastrophe theory is then to alert us to the possibilities of new behaviour and to look

for parameters with particular roles - such as the splitting
and normal parameters which were associated with the cusp catastrophe in Chapter 1.

We take a similiar stance in relation to differential
equations. While considerable insight into possibilities of
bifurcation can be gained from a qualitative analysis, our
ambition is always to seek to carry this through to building a
quantitative model. In any case, in applied work while it is
valuable to know the possible existence of critical points, it
is is often even more valuable to quantify them. The new theory,
however, does play a major role in teaching us what to look for
and how to put large models together. And there are many
instances where it enables short cuts to be made. In general,
however, the results of the main models presented below do not
depend on Thom's theorem in any direct sense - though they may
illustrate it.

The comment to be made about inductive versus deductive
approaches is of a similar character. In general, we are
involved mainly with deductive model building. But since we
are now alert to the possibilities of jumps, hysteresis and
divergence, as well as other kinds of bifurcation, then when
examples of such behaviour turn up in observations, we may be
encouraged to attempt to classify the transitions as a
preliminary to more formal model building.

3.6 A new focus for planning applications of models

The customary use of models in planning has been for
conditional forecasting: a set of values is assumed for some
controllable variables (and this set constitutes a plan) and
the models are used to make an assessment of impact. This
method will continue to be useful, especially for short and
medium-term forecasting. It will already be apparent, however,
that dynamical systems theory offers an additional and important focus: a concern with criticality and stability. If a
major change can take place in a system at some critical
parameter value, then it is obviously valuable to a planner to

know both that the possibility of this exists and also the actual value of the parameter which is critical. The knowledge may be important for two kinds of reason. First, the planner may wish to prevent the change taking place; secondly, he may wish to encourage it. In the first case he would be attempting to maintain the system in a stable state and not let 'conditions' (represented as parameter values) change in such a way that this disappears. In the second case, he may be trying to get the system into a new stable state by changing a parameter which may not obviously be connected with such transitions. Many examples of phenomena of this kind are presented in examples which follow. Two examples presented in a preliminary way in Chapter 1 also bear out the argument. One related to choice of transport mode, and in this case there may be a critical parameter value which determines the viability or otherwise of a public transport system. The second related to the possibility of development of facilities such as a shopping centre at a particular location. The methods of analysis offered here should ultimately give the planner the means to calculate the feasibility or otherwise of development at particular sites.

Note

(1) This notion of fast foliations off the manifold also connects intuitively to the topological representation of differential equations on a manifold. Given a surface of equilibrium values, the set of differential equations for all the variables involved is represented by a vector field on this manifold. In effect, it gives the speed of return to equilibrium following a disturbance in any tangential direction from the manifold.

CHAPTER 4

MACRO-SCALE APPLICATIONS

4.1 Introduction

This chapter is concerned with applications at the macro scale (rather than meso or micro, which follow in subsequent chapters in that order). This is mainly because systems treated at the macro scale appear simpler - certainly more so than the meso scale and perhaps also, micro - since the system is usually described by a very small number of variables. In the examples chosen below, most authors are concerned with the city as a whole (Amson, Mees, Isard, Papageorgiou) or with a system of cities and settlement patterns (Casti and Swain, though their discussion is limited to 'one centre at a time', and Wagstaff); Dendrinos is concerned with a part of a city - its slums - while Poston and Wilson are concerned with single centres within a city but this seems appropriate to this section because of the small number of variables used for the main idea and because it is similar in spirit to the Casti and Swain approach. The meso implications of the Poston and Wilson approach will be spelled out in the next chapter. We also note that the available examples are theoretical (though Mees' and Wagstaff's examples in historical geography are exceptions to this).

Catastrophe theory has made us aware of the possibilities of discrete or rapid change in systems, and it is straight-forward to find examples at this scale of what may be inter-preted as such change. Amson (1975) gives a useful list which includes: pre-urban clustering in a primitive dispersed society; the ten-fold enlargement of an already large city in one generation; the relatively instantaneous depopulation of a city's inner core after centuries of lively inhabitation; the reversals of income, social or ethnic zoning patterns and their counter-reversal later; the exchange of dominance roles between rival cities; the spontaneous formation of a conurbation from its constituent forms. There must be many more examples. Their existence encourages us to think of finding explanations (or at least descriptions) in terms of catastrophe theory.

The examples below are presented as separate case studies, but we begin with some comments to alert the reader to examine each author's response to certain general issues.

Since macro applications involve only a small number of variables, the choice of state and control variables is particularly important. A second important topic is the choice of potential function and the way in which the underlying mechanism of system change is represented in the model. A number of features emerge in the examples. First, there is considerable diversity in choice of variables for dealing with fairly similar problems. This is serious in that, if all the control variables used by the different authors really were independent, then they should be combined together and would generate catastrophes of a higher order than are discussed.

Secondly, it is relatively rare for authors to attempt to specify the potential function and the detailed mechanism. It is much more common to try to match observed behaviour against canonical behaviour, replying on Thom's theorem to argue that this indicates that a potential function does exist and that the equilibrium surface takes Thom's canonical form. This avoids the problem of transformations relating an actual system to the corresponding standard form. Only Amson (1974) tackles this issue. Other authors have to decide, on a general basis, which of the control variables function in which may - for example, in the cusp case, which to take as the 'normal' factor and which as the 'splitting' factor (although some adopt simple alternative formulations).

4.2 Amson (1974): catastrophic modes of urban growth

Amson's work on catastrophic modes of growth arises from earlier work on spatial structure (Amson, 1972-A, 1972-B, 1973) and the first part of 1974). There, he related the spatial distribution of densities (for a symmetric city and so as a function of distance from the centre, r) to measures of pressure: first a rental, p, which tended to pull people outwards towards cheaper rentals; and secondly a gravity-like attraction to people living 'inside' that location. This leads, on various assumptions. to equilibrium models of spatial structure. In order to find examples of catastrophic change, however, he neglects spatial structure and in effect replaces a measure of the benefit of the interaction with the inner city by a measure of 'opulence'. Thus, he characterises his city (or part of a city) by three variables: τ, the density, p, the rental and Ω, the opulence. The first of these is taken as the state variable and the second two as parameters or control variables. So, density is being 'explained' in terms of rental and opulence levels.

His method is interesting (and more in the spirit of some of the meso applications described in the next chapter). He does not go for a potential function directly, nor even, as is the case with most of the macro examples, for a direct representation of catastrophe geometry in canonical form (two control variables implies a cusp catastrophe and so on), but he seeks ways of specifying the equilibrium manifold of the three variables. Thus, the next step in the argument is to review briefly his four alternatives for 'urban laws':

His first possible law is:

$$p = K\Omega\tau \tag{4.1}$$

where K is a constant. This simply says that rental is proportional to density times opulence. This can also be written in the form

$$p \cdot \frac{1}{\tau} = K\Omega \tag{4.2}$$

which takes rental times average space occupied to be proportional to opulence.

This equation can then be modified by removing minimum personal space, β, from the term on the left hand side:

$$p(\frac{1}{\tau} - \beta) = K\Omega \qquad (4.3)$$

and the state equation can then be written in the form

$$p = \frac{K\Omega\tau}{1 - \beta\tau} \qquad (4.4)$$

The third possibility arises from a second possible modification of Equation (4.1). It is assumed that the rental will be reduced by an amount proportional to the density to allow for crowding effects. This is achieved by replacing $K\Omega$ by $K\Omega - \alpha\tau$ for some suitable constant α. Equation (4.1) then becomes

$$p = (K\Omega - \alpha\tau)\tau \qquad (4.5)$$

and Amson takes the standard form of this equation as

$$p + \alpha\tau^2 = K\Omega\tau \qquad (4.6)$$

The fourth possible law is obtained by applying the two modifications taking account of personal space and crowding simultaneously to give

$$p + \alpha\tau^2 = \frac{K\Omega\tau}{1 - \beta\tau} \qquad (4.7)$$

Physicists will recognise the progression in the argument from an ideal gas law through imperfect gas laws, to van der Waals' equation. It is in the context of this equation that discontinuous behaviour was recognised much earlier than the development of catastrophe theory as such and why this is some-times known as the Riemann-Hugoniot catastrophe (*cf.* Fowler, 1972).

Amson then goes on to analyse the dynamics of a system behaving according to these different laws. Recall now that τ is the state variable and p and Ω the control variables. In each case, we solve the equation for τ. For Equation (4.1),

$$\tau = \frac{p}{\Omega K} \tag{4.8}$$

which represents essentially smooth behaviour.

Equation (4.4) gives

$$\tau = \frac{p}{Bp + K\Omega} \tag{4.9}$$

and again, the behaviour is smooth.

Equation (4.6) is a quadratic equation in τ and the condition for real roots is:

$$K^2\Omega^2 > 4\alpha p \tag{4.10}$$

In the case $\alpha < 0$, there are two real roots, but only one of them is positive. More interesting (and in accord with the assumptions which gave rise to the equation) is the case $\alpha > 0$ where there are two positive roots. Call these A_L and A_S for largest and smallest respectively. Then the largest root is given by

$$A_L = \frac{K\Omega}{2\alpha} \left[1 + (1 - \frac{4\alpha p}{K^2\Omega^2})^{\frac{1}{2}} \right] \tag{4.11}$$

In this example, p and Ω can be taken as essentially co-varying, so that effectively there is one independent control parameter. In either case. A_L decreases as p or Ω decreases, and vice versa for A_S, until the two roots 'meet' at the critical point, given by equality in (4.10). If we take p as the main control variable, then this can be plotted roughly as on Figure 4.1, which we immediately recognise as the fold catastrophe. The lower curve has been plotted as dashed to show that these equilibrium states are unstable, while those on the upper curve are stable. This can be seen again by reference to the figure, showing the areas for which $\dot{\tau}$ are positive and negative. The reasoning for this is as follows: the differential equation for τ based on Equation (4.6) can be taken as

$$\dot{\tau} = K\Omega\tau - p - \alpha\tau^2 \tag{4.12}$$

71

Chapter 4

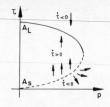

Figure 4.1 Amson's third law and the fold catastrophe

It is easy to check that this gives directions of change and return to equilibrium which are in accord with the original assumptions. This can be written in the form

$$\dot{\tau} = -(A_L - \tau)(A_S - \tau) \qquad (4.13)$$

using the definitions of the roots given earlier, and it is then an easy step to identify the signs of $\dot{t}$ on Figure 4.1 and to add the pieces of trajectory shown and hence confirm the above statement on stability.

This case shows that when certain parameters are reduced beyond a particular point, the density of the city becomes imaginary - which Amson calls 'evanescent'.

The fourth possible law is perhaps the most interesting. Amson shows explicitly that, by making transformations of his variables the equilibrium surface can be presented in standard cusp form. The transformations are

$$\tau = 1 + t \qquad (4.14)$$

$$p = 1 + q \qquad (4.15)$$

$$\Omega = 1 + w \qquad (4.16)$$

and with values of the constants as $\alpha = 9$, $\beta = \frac{1}{3}$, $K = \frac{8}{3}$, the equilibrium Equation (4.7) transforms to the cubic equation

$$3t^3 + (8w + q)t + 8w - 2q = 0 \qquad (4.17)$$

which is the familiar basis of the cusp catastrophe. The resulting surface (with the additional conditions $8w + q = y$, $8w - 2q = z$) is shown in Figure 4.2 and the transformations between the sets of variables on the control manifold in

72

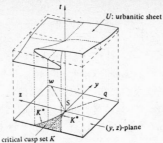

Figure 4.2 Amson's fourth law and the cusp catastrophe

Figure 4.3. Some typical paths around the control manifold are
shown on Figure 4.4. An interesting point here is that because
the control variables are transformed before the surface is
obtained in canonical form, then neither of the two original
variables can be taken as pure splitting (or normal) factors.
Also, it is possible, since the transformations are explicit, to
reconstruct the potential function. After some manipulation,
reversing all the transformations, this can be seen to be

$$v = \frac{3}{4}\tau^4 - 3\tau^3 + (8\Omega + p + 1)\tau^2 - 3(p + 1)\tau$$
$$+ (\frac{5}{2}p - 4\Omega + \frac{17}{4}) \tag{4.18}$$

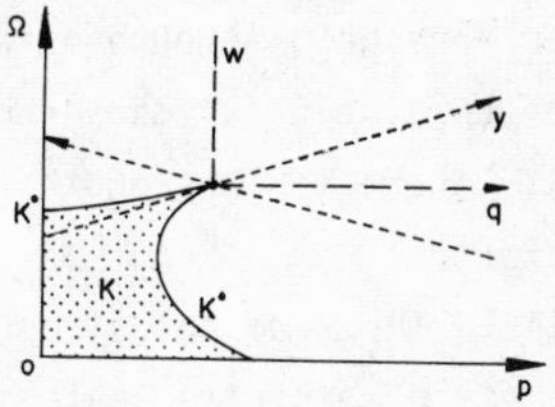

Figure 4.3 Amson's fourth law : the control manifold

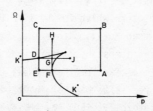

Figure 4.4 Amson's fourth law : typical paths in the control manifold

73

Amson calls this fourth example the *saccadic* city (and has thus
located in the dictionary a new word for jump behaviour). It
can easily be seen that crossing the bifurcation line in (p,τ)
space can lead to the usual kinds of jump behaviour in τ.
Because both factors contribute to the splitting function,
Figure 4.3 shows that a sufficiently high value of either
opulence or rent will ensure that all the changes are smooth.

4.3 Casti and Swain (1975) 1 : central place theory

These authors present two examples. The first is concerned
with the order of a central place in the usual sense. The
nature of the argument is the empirical identification of jump
and hysteresis behaviour of the system which suggests the
existence of an underlying catastrophe theory mechanism. This
therefore illustrates an inductive approach. The observed
phenomenon can be summarised as follows: if in some market area
there is a reduction in population or spending power, then there
exists a threshold, the lower threshold, below which goods of a
particular order will not be sold at the centre (which then
becomes of a lower order); if population or spending power
increase, however, such a change reverses itself not at the same
place, but at an upper threshold (though note this is different
from the use of 'upper threshold' as the 'range of a good' in
central place theory); there is thus another jump, but a
visible hysteresis effect.

To represent this in terms of catastrophe theory, they
take the functional level of a centre as their state variable
and two control variables - x, the population and y, the dis-
posable income per capita. This, they argue generates the cusp
catastrophe the the picture which is shown in Figure 4.5.

This diagram is used to illustrate the three main features
of catastrophe theory: jumps, hysteresis and divergence. We
can leave this as an exercise for the reader. The interesting
feature of their diagram is perhaps again the fact that they do
not distinguish one control variable as the splitting factor
and as the normal factor. This can be seen from the position

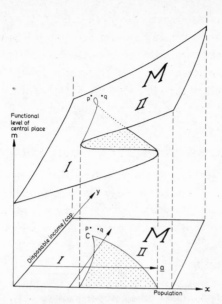

Figure 4.5 Casti and Swain : order of central places and the cusp catastrophe

of the cusp projection on the control manifold. However, they give no discussion as to why this particular shape has been assumed. This is an important research task if this sort of picture is to be pursued. It involves spelling out the different effects of the two control variables. As the argument stands at present, most if not all of the central place theory behaviour described at the outset could be ascribed to a single control variable - spending power - and the resulting picture would then be the fold catastrophe.

4.4 Casti and Swain (1975) 2 : property prices

Their second example is concerned with property prices. Here the state variable is r, the rate of change of property prices. They assume, first, two kinds of buyers: consumers and speculators and define D_c and D_s to be the demands for housing from the two groups. The resulting cusp picture is presented in Figure 4.6. Again, the cusp projection takes

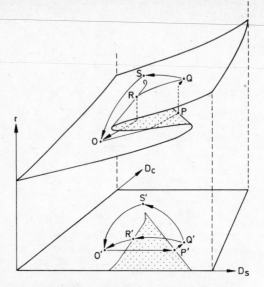

**Figure 4.6 Casti and Swain : urban property prices and
the cusp catastrophe**

rather the same form as it does in the previous figure, but
again with no detailed explanation of its form.

A more interesting illustration involves the addition of
new control variables. In particular, they recognise that a
government may try to use the interest rate to control the market and to avoid the most serious jumps. They do this by using
the butterfly catastrophe which enables us to explore a hitherto
uninvestigated part of Table 1.1. The potential function is

$$z = \frac{1}{6} x_1^6 + \frac{1}{4} u_1 x_1^4 + \frac{1}{3} u_2 x_1^3 + \frac{1}{2} u_3 x_1^2 + u_4 x_1 \qquad (4.19)$$

Casti and Swain's analysis of this catastrophe is a usefully
clear one and worth following in some detail. If u_1 is positive,
then for small x_1, the x_1^4 term dominates the x_1^6 term and the
surface is essentially the same as the cusp. The $u_2 x_1^3$ term acts
as a bias factor and changes the shape of the cusp, as shown on
Figure 4.7 for the $u_2 > 0$ case.

76

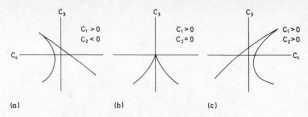

Figure 4.7   Sections of the (u_1, u_2) control manifold for the butterfly catastrophe : $u_1 > 0$

When u_1 is negative, the cusp divided into three cusps which form a pocket. The resulting projection on the $u_3 - u_4$ part of the control manifold is now shown in Figure 4.8. For this reason, the u_1 variable is known as the butterfly factor.

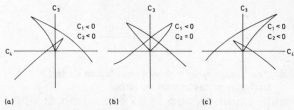

Figure 4.8   Sections of the (u_3, u_4) control manifold for the butterfly catastrophe : $u_1 < 0$

The argument now is to introduce interest as a control variable to ensure that the middle pocket of the butterfly surface is used to avoid big jumps in the market. This is done by taking the bias factor, u_2, to be the interest rate and the butterfly factor, u_1, to be the negative of time. In this way, it is argued, the intermediate surface is made available, and the scale of it (*cf.* Figure 4.8) is determined by the interest rate. The butterfly surface, for $u_2 = 0$ and negative u_1, is shown as Figure 4.9.

The problem with this argument is that there seems to be no rationale for using time as the fourth control variable in this case, and it is by no means clear that the intermediate surface does in fact exist in relation to the dynamics of this particular system. Thus, while this example gives us some useful insight into the potential of the butterfly catastrophe, it is by no means as yet a convincing one.

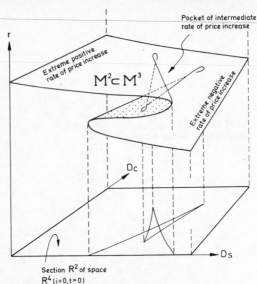

Figure 4.9 Casti and Swain : A representation of the butterfly catastrophe manifold

4.5 Poston and Wilson (1976) : another approach to centre size

This argument is essentially a geometrical one which involves the direct construction of a potential function which can then be shown to be interpretable in terms of the fold catastrophe. It is an illustration of how catastrophes can emerge from an attempt to study a mechanism in detail; and indeed this particular mechanism turns up in a number of fields. The example is taken as a macro one mainly because it involves only a small number of variables in its formulation and also because it complements the previous examples in a useful way.

As in the previous case, the topic is the size of some centre or facility - but in this case of one particular type. For the sake of fixing ideas, it is taken as a shopping centre, but the idea is more widely applicable in an obvious way to a range of urban and regional facilities. The basic idea is this: let W be the size of the centre and ℓ be the average distance a consumer has to travel to use it. It is clear intuitively that there is a monotonic relationship between W and ℓ in that the

larger W, the fewer centres there will be, and so the further
the consumer will have to travel. How can we decide what the
optimum (or the actual, if the determining process is an
optimising one) size of such centres is likely to be?

The argument is carried forward from the point of view of
the consumer. It is assumed that there are two components to
the consumer's utility: the benefits of facility size (u_1) and
the disbenefits of travel (u_2). The optimum size will result
from the trade off between these two elements. We assume that
u_1 increases rapidly with size at first, but is bounded, giving
it a logistic shape; and that travel disbenefits increase
linearly with distance travelled. Since ℓ and W are assumed to
co-vary, we work only with ℓ as dependent variable. Plots of
u_1 and u_2 against ℓ thus take the form shown in Figure 4.10.

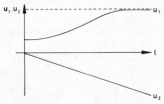

**Figure 4.10 Poston and Wilson : two components of utility
for shopping centres**

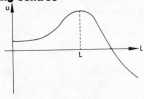

Figure 4.11 Poston and Wilson : total utility

The analysis now follows in a straightforward way.
Figure 4.11 shows the sum of the two curves (a plot of
$u = u_1 + u_2$), and we can assume that the optimum value of ℓ
(and hence, W) occurs when u is a maximum. We now need to
introduce a control parameter to see how the position of this
optimum varies with the parameter. The obvious one to choose is
the negative of the slope of the transport disbenefit line -

say, β. This can be taken as measuring 'ease of travel'.
Figure 4.12 now shows plots of u_1 and u_2 (the former not changing of course) and of u for a range of gradients - values of β.

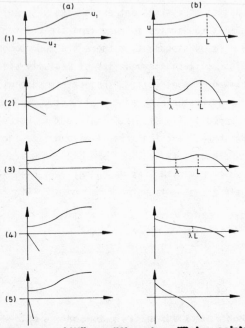

**Figure 4.12 Poston and Wilson: different equilibrium points for
changing 'ease of travel'**

The second column shows that for low β there is one optimum
value of ℓ but that a local optimum starts to emerge at the
origin as the slope increases.   For a range of values, as β
increases there are the two possible optima and eventually, the
origin becomes the global optimum.   Then beyond some further
point, the origin is the only optimum: there is a single
maximum of u again.

This information can be collected together in a different
way by plotting the values of the optimum values of ℓ against β.
This is done in Figure 4.13.   This shows that there are two
critical values of β.   The curve can be recognised as the fold
catastrophe curve, but with added states representing the

origin. This example of the fold is therefore not a pure one: the reader will recall from Chapter 1 that the fold does not have the zero states shown in Figure 4.13. These arise from the constraint that we must have $\ell > 0$ and so this can be taken as an example of a constraint catastrophe. It turns out to be quite common that additional state appear in fold catstrophe pictures through one mechanism or another - otherwise, the system would not have another state to jump to (as happened in Amson's third example above.)

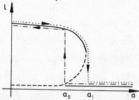

Figure 4.13 Poston and Wilson : shopping centre size and the fold catastrophe, with additional state

In this case, the origin solution is to be interpreted as the provision of a large number of small centres, while the other solution represents a small number of large centres. The model presents a picture of how one can get discrete jumps from one kind of solution to the other for small changes in the parameter β.

A number of possibilities for further work can be noted for this example. First, it would be easy to write down explicit functions for u_1 and, trivially, u_2 and hence to construct an explicit potential function u. This would not be in canonical form and would illustrate how transformations of variables could reduce it to such a form. However, it also emphasised how far it is possible to get with a simple geometrical argument. Secondly, it will immediately be noticeable that the control parameter used here is not used by Casti and Swain and vice versa - so there is some scope for increasing the number of parameters and thus increasing the order of the catastrophe surface which is relevant. On the other hand, we can also note that even if the real world actually does involve such parameters, the argument given above can be considered to present a slice of the higher dimensional case and is not invalidated.

Finally, we note that this example has meso-scale implications which will be taken up in Chapter 5.

4.6 Mees (1975) : the revival of cities in medieval Europe

Mees works rather in the style of Casti and Swain. He postulates a number of control variables and lets these determine the form of his analysis. His state variable is the population, p_m, working in manufacturing in towns and he has four control variables: τ, the difficulty of transport, q, the average productivity, δq, the different in productivity between town and country and C, a crowding factor (which is simply population divided by total land available). These definitions lead to the butterfly catastrophe. He takes q to be the butterfly factor and δq to be the bias factor. It is then possible to plot the (τ, C) part of the control manifold for various combinations of values of q and δq as shown in Figure 4.14. He traces the revival of cities as shown on parts 4 and 5 of the figure for alternative assumptions about the bias factor.

4.7 Isard (1977) : strategic elements of a theory of major structural change

Isard explores the potential contribution of catastrophe theory to theories of urban and regional structural change using the cusp catastrophe. He takes as his state variable the population x and as control variables α and β which represent the increase in productivity of each unit of population and the direct contribution of a marginal unit of population to total welfare, respectively. His potential function, W, is taken in canonical form as

$$W = -\frac{1}{4} x^4 + \frac{1}{2} \alpha x^2 + \beta x + C \qquad (4.20)$$

and this generates the cusp surface shown in Figure 4.15. It can be seen from the figure that α is the splitting factor and β the normal factor. The possibilities of different trajectories are illustrated on the figure.

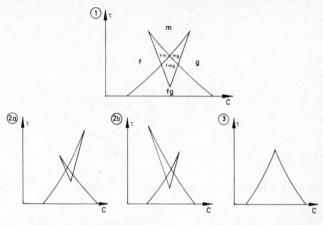

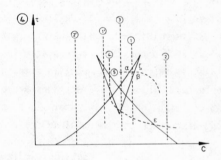

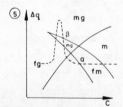

Figure 4.14 Mees : city growth and the butterfly catastrophe -
sections of control manifold and sample
trajectories

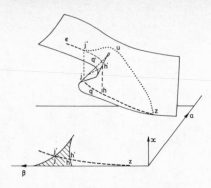

Figure 4.15 the cusp catastrophe as total welfare

Unusually, relative to the other examples, Isard attempts
to interpret the potential function directly (the other exception being Poston and Wilson above). It is always tempting to
take any function to be maximised as welfare, and this is what
he does. The βx term is taken as the direct contribution to
welfare and $\frac{1}{2}\alpha x^2$ as the positive agglomeration gains. The
$-x^4$ term is taken as representing negative externalities or
deglomeration forces. These interpretations are perhaps the
most important contribution of this paper. The paper also
includes a long discussion on the possible delay conventions to
be adopted in real cases, and this also identifies and
articulates a research problem of considerable importance.

4.8 Wagstaff (1978) : settlement pattern evolution

This author is concerned with a problem in historical
geography: the change in settlement pattern between the second
and seventeenth centuries in an area of Greece. He is able to
deal with space in a single dimension by drawing a line between
the extremes of his region as shown in Figure 4.16. There is
a shift in the positions of highest density from Malevri towards
the coast and there is some evidence that when this took place,
it was a rapid change.

This example is interesting as an illustration of unusual
choices of variables. Whether this is totally successful may

be a matter on which judgement should be reserved, but the ideas
are undoubtedly interesting and we shall concentrate here on
these methodological aspects.

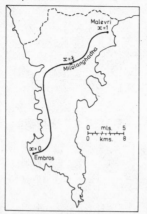

**Figure 4.16 Wagstaff : area of study, space co-ordinate as distance
along a curve**

The state variable of interest is essentially density, δ,
but the author takes this as the potential function, and so is
using the catastrophe theory framework to explain the relative
behaviour of this function.   As a state variable, he takes x,
the position along the curve shown in Figure 4.16.   The control
variables are T and F, the threat of attack and the quality of
agricultural land respectively. These are taken as simple
transformations of the canonical control variables and so are
connected to both normal and splitting functions.   This can be
seen from the potential function which is

$$\delta(X,T,F) \;=\; \frac{1}{4}\,x^4 - \frac{1}{2}\,(T + F)X^2 + (T - F)X \qquad (4.21)$$

Unfortunately, there is no detailed discussion in the paper
about the way the control factors operate and we have to confine
ourselves to the geometrical presentation.   The cusp catastrophe
model shown in Figure 4.17, which shows the T and F axes at 45^0
to the canonical control variables' axes. The possible path in
the control manifold which explains the density changes is shown
in Figure 4.18 and the section of the surface which shows the

85

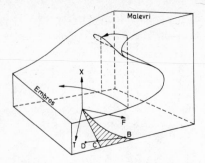

Figure 4.17 Wagstaff : the cusp catastrophe

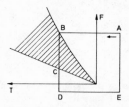

Figure 4.18 Wagstaff : possible trajectories on the control manifold

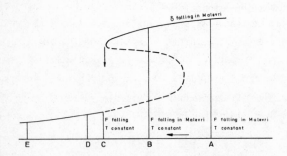

**Figure 4.19 Wagstaff : the potential function for changing
values of the control variables**

changes in δ is shown in Figure 4.19. The density changes
which are then being explained are shown in Figure 4.20.

(Note that this kind of plot would usually relate to the
state variable, in this case x, rather than the potential
function.)

86

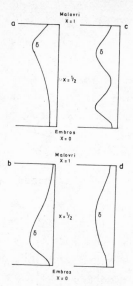

Figure 4.20 Wagstaff : settlement densities - taken as potential function

Wagstaff discusses the explanations which have been offered in the past for the change in settlement pattern. The main hypothesis involved an emphasis on external threat (T, above). Wagstaff, however, argues that the evidence is against this: while there were always external threats, the overall level is best considered as constant. On the other hand, there are possible reasons for deterioration in the quality of agricultural land (F) in the north: erosion and degradation which could have been caused by any one of a number of reasons, or a combination. His explanation, as sketched above therefore, generates the discontinuous change in density arising from a decrease in F for constant T.

87

Chapter 4

4.9 Dendrinos (1977) : slums in urban settings

Dendrinos' work is in the style of Casti and Swain and Mees in that he tries to identify the most important variables affecting the formation of slums at a macro scale and then to interpret the equilibrium state manifold and, briefly, the associated potential function. His state variables are x_1, the quality of housing stock and x_2, the utility level of the residents. There are four control variables: c_1, per capita income, c_2, the internal rate of return (which measures the willingness of private investors to invest), c_3, the social rate of discount (which measures the probability of investment by the public sector) and c_4, the population to capital stock density. This large array of variables then leads to a surface of higher dimension than any we have yet seen - the parabolic umbilic (sometimes known as the mushroom - which is given in Table 1.1 in Chapter 1) if his c's are replaced by our u's. The geometry is inevitably very complicated as shown by the various sections of the control manifold in Figure 4.21.

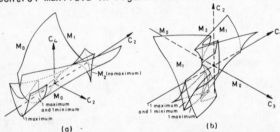

Figure 4.21 Dendrinos : urban slums and the mushroom catastrophe

He does not carry through his analyses of possible trajectories in any detail, but does make a number of unusual suggestions which can usefully be noted here. First, he suggests that the unstable states may in fact be achievable in the slum case, though he does not appear to give any strong evidence for this. Secondly, that in looking for possible system paths on the equilibrium manifold, it is a useful task to explore the density of such paths. And thirdly, he does attempt to interpret the potential function in terms of the minimisation of social costs.

4.10 Papageorgiou (1980) : sudden urban growth

Papageorgiou draws on empirical evidence that, at various times in history, many cities have grown suddenly in size and he constructs models to explain this phenomenon, building on earlier work by Wheaton (1974) and Casetti (1970). It turns out that his results, at least in part, can be described by the cusp catastrophe. The details of the model will not be presented here - only the main ideas in graphical form. The reader is referred to the original paper for details.

Let the population of the city be N, normalised so that when added to the total population in the alternative sector (say, rural), it is 1. That is, 1 - N can be taken as the population of the rural sector. Let $\bar{u}(N)$ be the average per capita utility achieved in a city of size N. Let $\bar{v}(1 - N)$ be a measure of utility of the alternative sector, which we will call 'rural' for brevity, when its population is 1 - N. The simplest possible assumptions about $\bar{v}$ and $\bar{u}$ are shown on Figure 4.22. $\bar{u}(N)$ is assumed to be a backward sloping straight line, and $\bar{v}(1 - N)$ also backward sloping line but drawn from right to left. They are equal at N = N*.

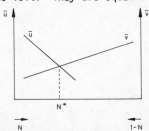

**Figure 4.22 Papageorgiou : urban and rural utilities –
no economies of scale**

It is assumed, therefore, that urban utility decreases as N increases and rural utility decreases as 1 - N increases. For N < N*, there will be rural to urban migration; for N > N*, urban to rural. This analysis also shows that the equilibrium point is stable.

In this model, it is assumed that technological change can shift both curves upwards over time. However, it is easy to see that any resulting change in N* is smooth if the upward movement of the lines is smooth.

Suppose, however, that we now allow for the scale economies of cities. Papageorgiou shows that $\overline{u}(N)$ can be backward sloping for low N, but then increase until it tails-off again as diseconomies set in. We assume that $\overline{v}$ retains its previous form. Appropriate plots are shown in Figure 4.24.

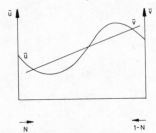

**Figure 4.23 Papageorgiou : urban and rural utilities –
with economies of scale**

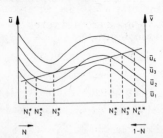

**Figure 4.24 Papageorgiou : effect of technological change
on urban size**

Consider now the effects of technological change. Suppose that the city benefits more from this than the rural area. Then the $\overline{u}$ curve will move up relative to the $\overline{v}$ line and, for convenience, we consider the latter to be fixed. A series of $\overline{u}$

curves are shown in Figure 4.24. Four cases are distinguished. The curve $\bar{u}_1$ has a single stable equilibrium point at a low value, N_1^*. $\bar{u}_2$ has two stable equilibrium points, a lower and an upper, at N_2^* and N_2^{**}. $\bar{u}_3$ is a limiting case: N_3^* is unstable from above; N_3^{**} is stable. At $\bar{u}_4$, the lower equilibrium point has disappeared, while the upper, N_4^{**} remains stable.

For a steady progression upwards of $\bar{u}$, the city will be initially in state N_1^* and will increase slowly to N_3^* (assuming something like a perfect delay convention holds). But as $\bar{u}$ increases further, the size must now jump from N_3^* to N_4^{**}, since at level 4 there is no lower equilibrium value. If we define τ to be the level of technology, we can plot N^* against τ as in Figure 4.25 and show the trajectory which has been described. The intermediate intersections, which are unstable states, are shown here as a dotted curve.

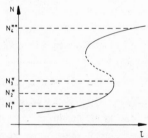

Figure 4.25 Urban change as represented by a section of the cusp catastrophe

This can clearly be interpreted as a section of a cusp catastrophe surface with τ as a 'normal' factor. Indeed, if we define s to be an index which measures the presence of scale economies, we see that the distinction between the $\bar{u}$ curve in Figures 4.23 and 4.24 was caused by s becoming positive, and that for s < 0, change in N^* is smooth. Hence s can be interpreted as a splitting factor and the whole cusp surface could be constructed by standard means.

Papageorgiou (1980) goes on to develop his model to incorporate more complicated forms of jump behaviour, but the reader is referred to the original paper for details.

4.11 Concluding comments

It is almost certain that the examples presented here
represent the beginning of a large number of such explorations.
Most of them are speculative and should be treated as such.
Perhaps therefore the main point to be taken in conclusion is
that a great variety of methods of approach are given even in a
relatively small set of examples. There are many choices of
control variables for similar problems, and this implies that
higher order catastrophes may someday be relevant. Meanwhile,
the examples presented could be seen as slices of such higher
order surfaces.

Perhaps more importantly, a variety of styles are
represented on how to treat and interpret the potential function, and how to treat the variables. These may roughly be
summarised as follows: (1) seek the equilibrium manifold
directly - as with Amson, Papageorgiou and, to some extent,
Poston and Wilson and then interpret the results in terms of
catastrophe theory; (2) define variables carefully, and then
use the canonical form of surface - Casti and Swain, Mees,
Isard, Wagstaff and Dendrinos. In this second category, we
have seen many ways of handling the variables within this
framework. The obvious thing to do is to take the state variables as the main dependent variables of the system of interest
(Casti and Swain, Mees, Isard and Dendrinos), but Wagstaff has
also shown that the main variable can be taken as the potential
function. It is also possible in some cases to take space and/
or time as control variables.

It is the second category of methods which are most
susceptible to critiques such as those of Zahler and Sussman
(1977). The equilibrium manifold is not constructed explicitly,
nor is any attention paid to the nature of the local region
within which Thom's theorem applies. This is why most of the
subsequent applications in this book are of the first type.
However, there will always be a difficult aggregation problem,
and applications at meso and micro scales will often have to be
linked to macro analyses.

CHAPTER 5

BIFURCATION AT THE MESO-SCALE I: COMPARATIVE STATICS OF URBAN SPATIAL STRUCTURE

5.1 Introduction

In the rapid development of urban models through the 1960s and into the 1970s, the most successful products have been those models which predict population activities on a given physical infrastructure (or which incorporate some very simple assumptions about such infrastructure being determined by 'demand'). As examples, we can cite models of transport flows, residential location and the use of service facilities, particularly shopping centres. These are partial models of the demand for the related elements of infrastructure - in the sense, at least, of being a response to supply. It has proved much more difficult to model the supply side. Such models, as they exist, still rely on the concepts of central place theory or something of that kind.

This position has often suited the modeller, especially when the main purpose of the model lies in its application in planning. The object of a planning exercise, typically, is to find some good or optimum location and size of facilities (transport networks, housing, shopping centres, hospitals, or whatever). Then, the models can be used to test the effectiveness of a particular plan. Such a plan would be a trial assignment of facilities (by size) to locations. These form exogenous inputs to models, and evaluation indicators can be constructed out of the model's prediction of population response to the plan. This remains a useful procedure in many instances. However, there are other cases where the modeller is functioning as analyst rather than planner and wishes to make the supply side endogenous to the models. This may be appropriate, for example, in the study of a city where there

93

is relatively little planning, or to build a picture of what an unplanned future would look like. There are also difficult intermediate cases, where planners have some control over some facilities, but not overall.

The next step in modelling, therefore, is to add mechanisms and equations to known models which determine supply side variables. We explore this in some detail below for the model of urban retail structure in various forms (the location and size of shopping centres as supply side variables), to residential location (and housing supply) and we couple them in an 'interacting fields' model, building on the well-known Lowry (1964) model. In the next section, we sketch the equations and their supply side extensions for these cases. Various possible theoretical bases of the models, based on mathematical programming, are mentioned briefly in Section 5.2.7; the details are presented in Appendix 1.

The next step in the argument is to show how to embed the extended models, which are essentially static equilibrium models, in a dynamic framework, either as differential equations or as difference equations and this is done in Section 5.3. We are then in a position, in Section 5.4, to use these equations to determine the stability of equilibrium points and to explore the nature of the possible equilibrium states. It is here that we first begin to see bifurcation properties which are relevant to a description of the evolution of urban structure and to get new insights from what remains essentially a comparative static analysis. A fuller exploration of dynamics is undertaken in Chapter 6.

5.2 The examples to be used
5.2.1 The urban retail structure model

The retailing system provides a good first example to illustrate the kinds of issues sketched in the introduction above. Consumers travel from residences to shopping centres (or from workplaces, an extension we will add below). The 'centres' are defined loosely at this stage to conform with the colloquial sense, but also to include small local shops if

appropriate. There are two elements to the modelling problem
as indicated above. First, there is the model of consumers'
behaviour: the response of consumers to a given spatial distribution of shopping centres. And secondly, the supply side:
the location and size of centres. This is the problem of
modelling the behaviour of suppliers, entrepreneurs or government planning agencies. They clearly would take into account
the spatial distribution of consumers and their spending power
in arriving at decisions on supply.

The first of the two modelling problems is the older one:
spatial interaction models of a now almost standard form have
been available to describe consumers' behaviour since the
early-1960s. The second problem is newer and is concerned with
the evolution of structure. The usual implicit hypothesis
about consumers behaviour is that there is a trade-off between
the benefits of centre size (cheaper prices, more choice and so
on) and the increasing costs of travel to larger centres - the
same kind of mechanism we saw in the Poston and Wilson example
in Chapter 4. This behaviour is represented in a spatial interaction model in which the location and size of shopping centres
is given exogenously. Such a model can then be used to predict
for example, the total revenues attracted to each centre.

To model the supply side explicitly, it is necessary to
make a hypothesis about how entrepreneurs (or other appropriate
agents) determine centre size and location. The one we will
use for illustrative purposes below involves the assumption
that they balance the costs of supply against the revenue
attracted at a particular location.

The spatial system and the associated variables are shown
on Figure 5.1. S_{ij} is the flow of cash from residents of
zone i to shops in zone j. (Two different zone systems, with
different numbers of zones in each if necessary, could be used
for residential areas and shopping centres - the latter
usually being considered as points.) e_i is the average per
capita expenditure on shopping goods by the residents of zone i.

Chapter 5

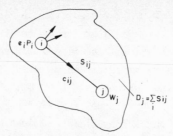

Figure 5.1 Flows to shopping centres

P_i is the population of zone i. W_j is usually defined as the attractiveness of shops in zone j and is commonly taken as being measured by the centre's size. Below, in Section 5.2.6, we discuss how composite attractiveness factors can be introduced, but this does not change the essence of the argument: that there is a term in the spatial interaction model of consumers' behaviour which measures centre attractiveness and which also serves as the structural variable - a measure of size at a location. c_{ij} is the cost of travel from i to j in suitable units - perhaps a generalised cost, a linearly weighted sum of travel time and money costs. In the model below, α and β are parameters and the A_i terms are the usual balancing factors.

The model was originally developed independently by Huff (1964) and Lakshmanan and Hansen (1965) and substantially developed by Harris (1965). An elementary account of it appears in Wilson (1974, Chapter 4). The simplest hypothesis is that the flows are proportional to the amount of cash available for spending at i, the attractiveness of shops at j and some decreasing function of travel cost. If we make some specific choices of the forms of the attractiveness function and the cost deterrence function, the model can be written

$$S_{ij} = A_i e_i P_i W_j^{\alpha} e^{-\beta c_{ij}} \tag{5.1}$$

where the proportionality or balancing factors, A_i, are given by

$$A_i = 1/\sum_k W_k^\alpha e^{-\beta c_{ik}} \tag{5.2}$$

to ensure that the set of constraints

$$\sum_j S_{ij} = e_i P_i \tag{5.3}$$

are satisfied. The model in this form can also be derived by entropy maximising methods (Wilson, 1970) or by any of a variety of other means as we will see in Section 5.2.7 and Appendix 1. The particular derivation is of no consequence for the main ideas to be presented below.

For a given set of shopping centre sizes, W_j, the model can be used to calculate the flows S_{ij} and hence the total revenue attracted to each centre (defined, say, as D_j):

$$D_j = \sum_i S_{ij} \tag{5.4}$$

The usual use of the model is to manipulate the W_j's by hand between model runs to seek an appropriate pattern. It is also possible to do this, as we will see in Appendix 1, within a formal mathematical programming optimisation procedure. However, we now seek an alternative hypothesis which will allow the W_j's to be determined endogenously.

If we assume that the suppliers of shopping centres balance cost and revenue, and that k is a suitable constant to describe the cost per unit of supply, then such a balancing condition can be written as

$$D_j = kW_j \tag{5.5}$$

for each zone j. Note that the number of equations added by (5.5) is equal to the number of unknowns W_j. Equations (5.1), (5.2) and (5.5) can then be treated as simultaneous equations in both S_{ij} and W_j variables. The non-linearities, mainly introduced through the nature of the A_i terms in Equation (5.2), ensure that the solutions have interesting bifurcation properties and we will be exploring these in Section 5.4 below.

5.2.2 Other models with a similar structure

The model described above can be seen as representative in a number of respects of other similar systems: people travel from home (at different average time intervals depending on the purpose of the trip) to use some facility or to take part in some activity - work, recreation, schools, health services, banks and so on. It is a relatively straightforward matter to construct analogous models. In Section 5.2.5 below, we consider the principles involved in disaggregating the shopping model, and this can also be considered to cover a variety of other services of the types mentioned. There is one rather special case, however, which is treated first and separately: that of residential location, which is often considered to be based on the journey to and from work. This gives the population behaviour side, if the model is assumed to have a similar spatial interaction structure to the retailing system, but there also turns out to be important differences. So, this case is considered first below. We then show how the retailing and residential model combine in Lowry-type models - 'interacting field' models - in Section 5.2.4. The remaining two subsections then deal with general questions of the forms of disaggregated models and composite attractiveness factors respectively.

5.2.3 Residential structure

Suppose people can be assumed to be located around a given supply of workplaces, E_j (which is employment in zone j). Then this can be taken as playing the role of $e_i P_i$ in the retail model and if we take W_i^{res} as a measure of attractiveness of i for residential location, then the simplest form of spatial interaction model which serves as a residential location model is

$$T_{ij} = B_j W_i^{res} E_j e^{-\mu c_{ij}} \qquad (5.6)$$

where

$$B_j = 1/\Sigma_k W_k^{res} e^{-\mu c_{kj}} \qquad (5.7)$$

to ensure that

$$\sum_i T_{ij} = E_j \tag{5.8}$$

The model now allows us to calculate total population allocated to each zone i (the equivalent of the revenue calculation in the retail model):

$$P_i = \sum_j T_{ij} \tag{5.9}$$

We have yet to specify the nature of the attractiveness function W_i^{res}. Working by analogy, we could take it as the number of houses available, say H_i, and then a suitable balancing condition for an equilibrium supply model would be

$$P_i = qH_i \tag{5.10}$$

for a suitable constant, q.

It is clearly an oversimplification to work by analogy. To begin with, the attractiveness function has to be more complicated, to represent other features such as environmental quality, density, nature of existing residents and so on. It is also likely that this will make sense in terms of a disaggregated model which distinguishes both person types and house types. There are also inertias associated with the housing stock and person movement within it which are both likely to be greater than those associated with shopping facilities and this too needs to be taken into account in an adequate dynamic model. We postpone further discussion, therefore, until we have outlined general principles for disaggregation and for the construction of composite attractiveness factors. The model as it stands above will, however, serve to illustrate the notion of interacting fields in the next subsection.

5.2.4 Interacting fields : the Lowry model

The model developed by Lowry (1964) has had an immense influence. It has many features which have not adequately been picked up in most urban models since its development - for example, the representation of shopping and service trips from workplaces; an attempt, albeit a simple one, to link

Chapter 5

demographic and economic assumptions with spatial location models; and a systematic treatment of land use accounting. However, there is one feature which is particularly striking and relevant for present purposes and on which we concentrate now: the way in which the two spatial interaction models which are at the heart of Lowry-style models are linked. They can be said to represent interacting fields of influence.

Suppose, for illustrative purposes, that the retail model given by Equations (5.1) and (5.2) represents the whole of the service sector and that the residential model given by (5.6) and (5.7) is also adequate as the basis for discussion of the simplest possible kind of Lowry model. The retail model can be considered to represent a field around population zones as indicated symbolically in Figure 5.1. The residential location model is based on another field of influence concept - this time around the given number of jobs. This is outlined in Figure 5.2. It is immediately clear that these two submodels

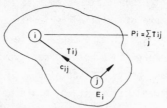

Figure 5.2 Flows to workplaces

are strongly coupled, and the greatest power of Lowry's work was to recognise this. The retail field, around population centres, generates retail facilities and hence the service employment at those points. This could be assumed to be given by

$$E_j^R = \gamma \sum_j S_{ij} = \gamma D_j = \gamma k W_j \qquad (5.11)$$

where the superscript R represents retail and γ is a suitable constant. Total employment is given by

$$E_j = E_j^B + E_j^R \qquad (5.12)$$

where the superscript B denotes basic (non-service) employment
which is assumed to be given exogenously. But this employment
generates the population distribution which in turn generates
the service employment, and so on. The resulting mathematical
problem can always be solved iteratively, and converges
straightforwardly. A pictorial representation of the field
interactions involved is given in Figure 5.3.

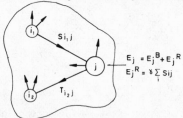

Figure 5.3 Interactions of two fields

Equations (5.1), (5.2), (5.6) and (5.7) together with
(5.11) and (5.12) form the heart of a Lowry model in its usual
form. Since E_j^R is proportional to W_j, the kind of additional
hypotheses discussed above which determine W_j also determine
E_j^R. Thus, Equation (5.5) could be added. Similarly,
balancing hypotheses of the form (5.10) determine P_i, and thus
also could be added. Later, when we have developed suitable
apparatus for the study of equilibrium behaviour for the two
models separately, we will therefore have the more complicated
case where the two models interact, to consider.

5.2.5 Disaggregation

We explain different principles of disaggregation in
relation to a number of examples. Our greatest interest here
is not simply on the obvious need to generate more realistic
models, but on the way in which new non-linearities and new
couplings between model equations are introduced through disaggregation. This, as we can expect from the general
discussions in Chapters 1 and 2, will lead to interesting new
bifurcation properties.

Chapter 5

Consider first the retail model given by Equations (5.1) and (5.2) and let us distinguish different types of goods, particularly with a view to representing some kind of hierarchical structure. The simplest way to proceed is to put a new label, g, for type of good (or service), throughout the equations:

$$S_{ij}^g = A_i^g e_i^g P_i \hat{W}_j^g e^{-\beta^g c_{ij}} \tag{5.13}$$

with

$$A_i^g = 1/\sum_k \hat{W}_k^g e^{-\beta^g c_{ik}} \tag{5.14}$$

to ensure that

$$\sum_j S_{ij}^g = e_i^g P_i \tag{5.15}$$

where $\hat{W}_j^g$ is the attractiveness term for good g to be used in model. Because expenditure on different types of goods has been distinguished, there is no immediate coupling of these equations. However, this can be achieved through an examination of the attractiveness terms $\hat{W}_j^g$. It is likely that the attractiveness of a shopping centre will depend in part on the overall size of the centre (say given by W_j^*, where we use the asterisk to denote summation over an index, and assume that the W_j^g's for different g are measured in comparable units), and partly on the provision for that particular good, W_j^g. Thus we might take

$$W_j^g = (W_j^*)^{\alpha_1^g} W_j^{g \alpha_2^g} \tag{5.16}$$

where we now introduce appropriate sets of parameters. Because W_j^* involves the W_j^g's at all levels, the form of attractiveness function given by Equation (5.16) clearly couples the different equastions. Alternative forms of attractiveness function can also be constructed. For example, if g represents levels in a hierarchy, then the first term may only involve elements of W_j^g for g's at higher levels in the hierarchy. This could be written

$$W_j^g = (\sum_{g'>g} W_j^{g'})^{\alpha_1^g} W_j^{g \alpha_2^g} \tag{5.17}$$

(for a suitable ordering of g-values to represent the hierarchy) so that the range of the first summation is defined to represent such an effect. The particular form to be adopted would depend on the results of empirical work. Here, we are interested in exploring whether different bifurcation phenomena arise from different forms of attractiveness functions arising in this kind of disaggregation. This in itself could provide clues for phenomena to search for in empirical work.

It remains, for this first example, only to note that the supply side equations which would be equivalent to (5.5) would be

$$D_j^g = k^g W_j^g \tag{5.18}$$

The position of goods and services in the hierarchy would be determined by the relative values of the α, β and k parameters.

The next step in the argument is to follow through in the notation of this book Lowry's recognition that trips to service or retail facilities were often made from workplace as well as from residential bases. Using an obvious extension of the notation developed so far, if we use the superscript (1) to denote trips from home and (2) to denote trips from work, we have

$$S_{ij}^{(1)g} = A_i^{(1)g} e_i^{(1)g} P_i \hat{W}_j^{(1)g} e^{-\beta^{(1)g} c_{ij}} \tag{5.19}$$

where

$$A_i^{(1)g} = 1/\sum_k \hat{W}_k^{(1)g} e^{-\beta^{(1)g} c_{ik}} \tag{5.20}$$

to ensure that

$$\sum_j S_{ij}^{(1)g} = e_i^{(1)g} P_i \tag{5.21}$$

and

$$S_{ij}^{(2)g} = A_i^{(2)g} e_i^{(2)g} E_i \hat{W}_j^{(2)g} e^{-\beta^{(2)g} c_{ij}} \tag{5.22}$$

where

$$A_i^{(2)g} = 1/\sum_k \hat{W}_k^{(2)g} e^{-\beta^{(2)g} c_{ij}} \tag{5.23}$$

to ensure that

$$\sum_j S_{ij}^{(2)g} = e_i^{(2)g} E_i \qquad (5.24)$$

Again, the main interest in such a scheme lies in the new kinds of couplings and non-linearities which are introduced. The latter arise from the different parameter values for the different types of trip (by home or workplace base). The former arise because the supply side equations which ultimately determine the W_j^g's will be as follows, and do not distinguish type of base - only total revenue attracted:

$$D_j^g = D_j^{(1)g} + D_j^{(2)g} = kW_j^g \qquad (2.25)$$

We noted in subsection 5.2.3 above that the residential location model needed to be disaggregated by both person type (say using an index w) and house type (say using k). A typical example of such a model would be

$$T_{ij}^{kw} = B_j^w H_i^k E_j^w e^{-\beta c_{ij}} \mu b_{ij}^{kw} \qquad (5.26)$$

This follows the argument of Wilson (1970) and Senior and Wilson (1974), though once again, it is simply being taken as a typical example of a disaggregated model and others could be substituted without affecting the essential argument about dynamics to be presented later. In this model, T_{ij}^{kw} is the number of type w people resident in a type k house in zone i and working in zone j. (We neglect here all the discussions which are needed to build in non-workers, households with varying numbers of workers and so on. For a more general review of residential location, see Senior, 1973, 1974.)

H_i^k is the number of type k houses in i. E_j^w is the number of type w jobs in j (person type being assumed to derive from job type - for example in relation to income). b_{ij}^{kw} is the bid rent for a type k house in i by a w-type person working in j. The other terms are as before, or are obviously parameters. An equilibrium condition based on this model would take the form

$$\sum_{jw} T_{ij}^{kw} = q^k H_i^k \tag{5.27}$$

for a suitable constant, q^k. The different equations in (5.26) are coupled in the usual way through the nature of the balancing factors, which in this case are

$$B_j^w = 1 / \sum_{i'k'} H_{i'}^{k'} e^{-\beta c_{i'j}} \mu b_{i'j}^{k'w} \tag{5.28}$$

to ensure that

$$\sum_{ik} T_{ij}^{kw} = E_j^w \tag{5.29}$$

(The model has been presented in singly constrained form, which is more useful as the basis for dynamical analysis, though this particular model is often presented in doubly constrained form.)[1]

The next obvious step in the argument is to investigate how the two sets of diaggregated models would combine together in a Lowry-like structure. As formulated, the most obvious connection would be to make per capita expenditure terms like e_i^g, dependent on the population mix in a particular zone, which is available as

$$P_i^w = \sum_{jk} T_{ij}^{kw} \tag{5.30}$$

using an obvious notation. The other features of linkage turn on a more detailed discussion of attractiveness factors and so this is postponed until the next subsection.

5.2.6 Composite atractiveness factors

What we think of as attractiveness or deterrence factors usually appear in spatial interaction models as power or as exponential terms. For example, in Equation (5.1) we have the terms W_j^α and $e^{-\beta c_{ij}}$. The entropy maximising or random utility derivations of the model show that the term in the exponential function acts like a cost or utility, and that a power function term can be similarly interpreted if written, as in the case of the example just considered, as $e^{\log W_j^\alpha}$. If we want the interpretation to be in money units, we can write the whole expression relative to such a term - c_{ij} for example. Thus:

$$W_j^\alpha e^{-\beta c_{ij}} = e^{-\beta\left(\frac{\alpha}{\beta} \log W_j - c_{ij}\right)} \qquad (5.31)$$

We see now the benefits of centre size are supposed to increase logarithmically with size, which is quite a sensible hypothesis.

The main point of this digression on the nature of attractiveness and deterrence functions is to note that any property which is thought to be relevant to the corresponding decision process represented in the model can be built in a straightforward way. In particular, attractiveness terms can easily be composite. For example, the residential attractiveness term, W_i^{res} may take the form:

$$W_i^{res} = X_{1i}^{\gamma 1} X_{2i}^{\gamma 2} X_{3i}^{\gamma 3} \ldots \qquad (5.32)$$

where the X_k's are the multiplicative components and the γ_k's suitable parameters. In this way, very rich behaviour can be represented in the models. More importantly for our present purposes, this leads to new non-linearities and new couplings being incorporated. The non-linearities arise in an obvious way because of the nature of the parameters γ_k. The couplings arise because some of the X-terms may be derived from related models, or even from the model itself, thus building in direct feedback. As illustrations of each of these concepts consider the following. Residential attractiveness may be a function of accessibility to shops. Such a quantity can easily be computed using terms from the retail model and then built in as one of the X-factors. Secondly, attractiveness may be a function of residential density: this is population divided by land use, and so one of the X-factors can depend, in a non-linear fashion, on one of the terms, the population in the zone, which is being computed directly by the model. These are all likely to be the basis of interesting bifurcation properties.

A converse linkage was hinted at in the previous subsection: to make retail attractiveness a function of the

quality of goods on supply which is likely to be related to the
wealth of the population using the shop.   An index of this
could be taken as the potential

$$\sum_{i:e_i>\bar{e}} e_i P_i e^{-\beta c_{ij}} \tag{5.33}$$

(where we introduce $\bar{e}$, in effect, as a new parameter) for
example, and this could be built in as a term in the attractiveness function and raised to an appropriate power.

Most of the further discussion of attractiveness factors
will be left until we get involved with more realistic dynamic
models in the various sectors later. Two or three more points
can usefully be made at this stage. First, when the attractiveness factor is composite, usually there will be only one
factor which is the subject of a balancing condition such as
(5.5) or (5.10).   It is important to distinguish this carefully
and also to define the equivalent equation carefully but the
essence of the argument does not change.

Secondly, there is sometimes a choice within the model as
to how to build in a particular factor; or, to put this
another way, a particular factor may affect either the demand
side or the supply side or both (in proportions which can only
be discovered from very difficult empirical research).   For
example, it can usually be reasonably assumed that availability
of land is not an important factor for retail location, but
this will certainly not be the case for residential location.
It may affect the demand side through a term in the attractiveness function which reflected density preferences, and
obviously after a time, density would become unacceptably high
as land ran out. It might also be considered as a supply side
factor, affecting terms like q in Equation (5.10). This would
be made a function of population allocated and land available
and would start to become very high as land ran out. We will
see the effect of this when we begin our dynamical analysis
from Section 5.4 below and in Chapter 6.

5.2.7 Theoretical foundations of the models : mathematical programming

Through this section, we have presented the basic range of models to be used, but without giving any detailed derivation of them. We presented briefly a 'common sense' argument in terms of proportionality and indicated that entropy maximising and other derivations were possible. Most of these alternative derivations can be presented as mathematical programming formulations and in this subsection we summarise the variety available. They are presented in detail in Appendix 1. The alternative derivations provide a number of different insights. They help with the interpretation of various terms within the models. Also, alternative methods are provided for the solution of the equation system (such as (5.1), (5.2) and (5.10)) for the equilibrium points. Further, general theorems of mathematical programming can be invoked to say something about the existence and uniqueness of equilibrium points. We develop and use these results in Section 5.4 below. Finally, since the models can be set up in Lagrangian form as unconstrained problems, these formulations also provide an alternative basis for the development of differential equations - so-called Lagrangian dynamics.

In Appendix 1, the first alternative to be discussed is the entropy maximising formulation, using the shopping model as an example (A1.2). In subsections A1.3 and A1.4 various aspects of consumers' surplus are explored and in the second of these sections, we incorporate an important discussion on 'embedding'. Then, accessibility maximisation and random utility theory are discussed in turn. We note in Section A1.7 that the application to residential location is straightforward, but this is less so for the Lowry model and the basic results for this case are presented in Section A1.8.

5.2.8 Summary of the position reached

It is perhaps best to begin a summary by taking a step back and reviewing the models which have been considered in the light of a comprehensive model system: how many components of such a system have been comprehensively dealt with? Consider

108

the sketch of the whole model system shown in Figure 5.4
(following Wilson, 1974, for example, and earlier papers such
as Wilson, 1968).   This broadly divides cities into subsystems
concerned with demographic and economic development on the left
and right hand sides of the figure, and shows population and
economic activities, spatially distributed, connected to the
appropriate components.   Physical infrastructure and transport
flows are shown arising from (though of course also partly
creating) these distributions of activities.   The demographic
and economic models - the latter considered as a function of
assumptions about technological advance - are considered as
parameters, as driving terms, in relation to urban development.
They will trigger bifurcations in other submodels, but we do
not investigate bifurcation properties of these submodels
directly.

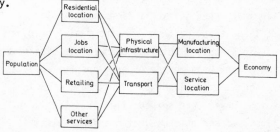

Figure 5.4 The main urban subsystems

The population activity models are those which have been
available since the 1960s and have been reviewed here.   We are
now arguing, however, that we can extend these models so as to
begin to model the distribution of physical infrastructure, the
subsystem shown in the centre of the diagram (with the exception as yet of dealing with the evolution of transport networks)
and, in the same way, the supply of economic activity in the
service sector (which may be well over 50 per cent of an
advanced economy).   The remaining gap, therefore, is the spatial
distribution of basic economic activity.

The aim of this section has been to present the principles
involved in the basic models to be considered in detail in the

rest of this chapter. By showing how they can be disaggregated and coupled in various ways, we hope to have indicated how they can be made appropriately realistic. The discussion of composite attractiveness factors was also intended as such a contribution. In the following sections, however, we will use the simplest version of any particular model with which it is possible to illustrate a new methodological advance. Then later, in Chapter 6, we collect together the advances in dynamic modelling with the task of adding greater realism, and consider integrated approaches - towards a new central place theory?

5.3 Equations for dynamical analysis

5.3.1 Introduction

In this section, we build on the model developments reported in the preceding two sections by articulating the differential equations which provide the basis for dynamical analysis. This applies both to 'comparative static' analyses, when the equations provide the foundations for analyses of stability, and to disequilibrium situations, when the differential equations have to be solved directly. First, in Section 5.3.2, we set up the equations for the retailing example and demonstrate the general principles involved. We show how to present equivalent equations in difference equation form in Section 5.3.3. In the next three subsections, we then show briefly how to apply the principles to the other models considered earlier: the residential model, the Lowry model and various disaggregated models respectively.

5.3.2 Retail model differential equations

We have now given a detailed presentation of the basic model involving urban structural variables and the next step in the argument is to seek differential equations which are the basis of their dynamical behaviour. We saw in Section 5.1.1 at the beginning of this chapter that new equations had to be added to determine equilibrium values of variables such as W_j

and we now seek to establish what happens when the system is disturbed from that equilibrium. This also helps us to analyse the stability of equilibrium states. The additional balancing equations considered in Section 5.1.1 are repeated here for convenience:

$$D_j = kW_j \qquad (5.34)$$

The simplest possible differential equation for W_j is one which says that W_j grows if $D_j - kW_j$ is positive and declines if it is negative. For a suitable constant ε, the simplest such equation is

$$\frac{dW_j}{dt} = \dot{W}_j = \varepsilon(D_j - kW_j) \qquad (5.35)$$

This can be recognised as one of the growth equations discussed in Section 2.4.4 of Chapter 2. D_j is not, of course, a constant and it can be written out in full, using Equations (5.1), (5.2) and (5.4) from the beginning of the chapter

$$D_j = \sum_i \frac{e_i P_i W_j^\alpha e^{-\beta c_{ij}}}{\sum_j W_j^\alpha e^{-\beta c_{ij}}} \qquad (5.36)$$

so that the differential equation (5.35) can be written in full (to emphasise their non-linear character) as

$$\dot{W}_j = \varepsilon \left[\sum_i \frac{e_i P_i W_j^\alpha e^{-\beta c_{ij}}}{\sum_j W_j^\alpha e^{-\beta c_{ij}}} - kW_j \right] \qquad (5.37)$$

It will also sometimes be convenient to write the equations in the more general form

$$\dot{W}_j = M_j(W_1, W_2, \ldots, W_N)W_j \qquad (5.38)$$

for obvious definitions of M_j for this particular case. We examine the general properties of these equations, in an ecological analogy, in Section 5.4.5.

Chapter 5

It is useful to compare the form of equations derived above with those used in the general discussion of growth equations in Section 2.4.4 of Chapter 2. The system of equations is rather like that in (2.33), with the feedback generated by all the D_j's being non-linear functions of the W_j's, but with a W_j factor missing. Later in that section we did indeed explore the whole family of growth equations, which stated for a single variable, would be

$$\dot{x} = \epsilon x^n (D - x) \qquad (5.39)$$

where n is a parameter. In the example above, n = 0; for the logistic form of growth (for the variable as a function of time), n takes the value 1. For n < 0, the gradient of the pick up near the origin is steeper, and indeed infinite in the limit; for n > 1, the form of the inflexion in the logistic plot becomes more pronounced. This suggest that we can add a factor W_j^n to the differential equations given earlier (that is, in effect, to take a value of n other than zero) and change the shape of the pick-up of W_j as a function of time without changing the essentials of the dynamic behaviour: any bifurcation and stability considerations associated with equilibrium points will always turn on the term in square brackets in Equation (5.37) as we will see shortly.

One particular point is an interesting one to bear in mind: if we think of the system as a gradient system, where we are maximising the Lagrangian function given in Equation (A1.12) of Appendix 1 but with an additional term to account for the constraint (A1.19) which has to be added when we have W_j variations as here, then the usual differential equations which are first considered in dynamical systems theory are

$$\dot{W}_j = \frac{\partial L}{\partial W_j} \qquad (5.40)$$

(or with a minus sign if the problem is a minimisation one). In this case, the Lagrangian is

112

$$L = -\sum_{ij} S_{ij} \log S_{ij} + \sum_{i} \lambda_i (e_i P_i - \sum_{j} S_{ij})$$

$$+ \alpha (\sum_{ij} S_{ij} \log W_j - \overline{\log W_j})$$

$$+ \beta(C - \sum_{i} S_{ij} c_{ij}) + \gamma(W - \sum_{j} W_j) \qquad (5.41)$$

Some manipulation shows that the form which the equations (5.40) take is the same as (5.37) but with a term $1/W_j$ added to each equation in turn. In other words, they take the same form as the general equations with n = -1, which gives an infinite gradient pick-up at the origin when the growth of W_j from an initial disequilibrium position is plotted as a function of time. It is possible to give a Lagrangian which generates (5.37) exactly, but this turns out to be not easy to interpret. This point is discussed in Appendix 2.

Allen *et al.* (1978) tend to use the logistic form of growth equation and to have a slightly different and more general expression for D_j: for example

$$\dot{W}_j = \varepsilon W_j (D \frac{F_j}{\sum_k F_k} - W_j) \qquad (5.42)$$

(in the notation of this chapter). Here, D is total demand for the service and $DF_j/\sum_k F_k$ is the proportion attracted to the centre at j. The main description of the use of Allen's forms of equation will come in the section on disequilibrium in Chapter 6. The other equations will be used in a variety of ways in the following subsections, first to help in the analysis of the stability properties of equilibrium points in Section 5.4.2, following a brief note on the treatment of similar equations in difference equation form.

5.3.3 Difference equations

It is a straightforward matter, of course, to rewrite any differential equations introduced so far as difference equations, and indeed this will often be necessary for computational reasons. The Equations (5.35) could be written

$$W_{jt+1} - W_{jt} = \Delta t . \varepsilon (D_j - kW_j) \qquad (5.43)$$

White (1977), for example, has carried out much computation of equilibrium points in this format. However, we will not pursue the point any further (until Chapter 8) except to note that, in the literature, there are equations in difference form which may not immediately be recognisable as taking the form of the equations we are working with and therefore having similar bifurcation properties. One example is a simple model of employment location used by Putman and Ducca (1978). This takes the form (using an obvious notation):

$$E_{jt+1} = \lambda \Sigma_i P_i \frac{W_j f(c_{ij})}{\Sigma_k W_k f(c_{ik})} + (1 - \lambda)E_{jt} \qquad (5.44)$$

and can clearly be written as

$$E_{jt+1} - E_{jt} = \lambda \left[\Sigma_i P_i \frac{W_j f(c_{ij})}{\Sigma_k W_k f(c_{ik})} - E_{jt} \right] \qquad (5.45)$$

which is basically the same form as the retail model though, of course, with the parameters in a different form and playing different roles. Indeed, in this case, the λ parameter is more like the ε parameter and, in effect, the value of k (if the equation is taken as an analogue of (5.35)) is 1. However, this does not prevent the equations having interesting properties as we shall see later. It also provides an interesting extension of the range of basic models, since employment here can include basic employment.

5.3.4 Residential location

The important point to note about the retail example in 5.3.2 above is that the structural differential equation arises out of the balancing condition. In the residential location case, therefore, we look to Equation (5.10). An analogous equation for $\dot{H}_i$ would then be

$$\dot{H}_i = \rho(P_i - qH_i) \tag{5.46}$$

where ρ is an additional constant. The same argument about alternative forms of the equation arising from a factor H_i^n on the right hand side apply as in the retailing case. P_i would be obtained from Equation (5.9), T_{ij} from (5.6) and B_j from (5.7) and this leads to a set of non-linear simultaneous equations which will have the same (in a formal sense) bifurcation properties as the retail model. Of course, in this case, the situation will be made much more complicated by the introduction of composite attractiveness terms along the lines suggested in Section 5.2.6. That is, if W_i^{res} in (5.6) is taken to be of the form (5.32), housing availability, H_i, will be one factor in that expression and other terms may themselves appear in other differential equations (for example, to take an obvious illustration, retail centres if an 'access to shops' term is included). This will obviously make the bifurcation behaviour more complicated. We shall consider specific examples of this model, and composite attractiveness factors, in Chapter 6 when we try to draw all the threads of the argument together.

We noted in the initial discussion on residential models that there were two separate questions of inertia: that of the housing stock and that of the population using it. If we concentrate only on the housing differential Equation (5.46) above, then this implies that the population move into equilibrium reasonably rapidly. If we wanted to vary this assumption, this would involve the development of a separate differential equation for population by zone. We note later, in discussing the work of Allen *et al*. (1978) that other authors in fact develop differential equations directly for population. This involves the other kind of assumption that housing always 'follows' population and appears accordingly.

5.3.5 The Lowry model

For the present, we will simply assume that both sets of differential equations, for retailing and for residences, developed above can be separately added to any dynamic Lowry model. A more complicated alternative argument is possible based on Lagrangian considerations, but this will divert us from the main point of the present argument, and the reader is referred to the paper by Macgill and Wilson (1979) for more details of this.

5.3.6 Disaggregated models

Again, at this stage, we simply need to observe that the differential equations are based on the balancing conditions such as (5.18) or (5.25). In the first case, the equation would be

$$\dot{W}_j^g = \epsilon^g (D_j^g - k W_j^g) \tag{5.47}$$

and the bifurcation properties would arise from the complexities in the attractiveness factors - as in Equations (5.16) or (5.17). In the case of (5.25), the equations are the same, but there will be new bifurcation complexities because of the way D_j^g is computed.

5.4 Equilibrium point analysis

5.4.1 Introduction

In this section, we explore the nature of the equilibrium points of the model we have set up for the prediction of structural variables. In the retail case, for example, we focus on the pattern of W_j-variables which occur when $\dot{W}_j$ is zero and investigate the bifurcation properties of these patterns as various parameters change. For the purposes of such an exercise, any variable which is exogenous for the particular problem has to be treated as a parameter: since it can change, it can have critical values. Thus, while we will focus initially on parameters such as k, α and β, we will later treat terms such as e_i and P_i as parameters.

We are mostly taking a rather narrow view of bifurcation theory: a concern with changing equilibrium points, and a focus particularly on critical parameter values at which such points appear or disappear. We do not, at this stage, investigate the possibility of transitions to other forms of solution. Nor do we make much direct use of catastrophe theory. We only use it to indicate the kinds of singularities which might exist in the surface of possible equilibrium states (and of course we are using it in the sense discussed in Chapter 1, that we are made aware of the possibility of unusual kinds of behaviour occuring).

The methods to be explored therefore, are essentially those of comparative statics. Indeed, it could be argued (Harris, private communication) that developments in this kind of bifurcation theory should lead to a resurrection of interest in comparative statics: in the past, it has typically been assumed that the transition from one equilibrium state to another as a parameter changes smoothly will be smooth, but we now know this is not necessarily the case.

Most of the argument is cast in terms of the retail model but then, in subsequent subsections, we note any extensions or differences that arise with our other examples: residential location, the Lowry model and with disaggregated models. We also add a section on the insights to be gained from examining the differential equations from an ecological perspective. In the final two subsections, we summarise the possible kinds of jump behaviour and discuss the difficulties of interpretation which this creates, and then we present some numerical illustrations.

Chapter 5

We mentioned in Section 5.2.7 and Appendix 1 that mathematical programming procedures might help us to find equilibrium points. There is, in fact, a range of possible algorithms for calculating these points. Here and below, we concentrate only on the theory and the reader is referred to Harris and Wilson (1978) and Phiri (1979, 1980) for a review of alternative methods.

5.4.2 Retail model equilibrium point theory

The differential equations for the rates of change of the structural variables, W_j, representing the spatial pattern of shopping centres and their sizes, were given as (5.35) in the previous section. The equilibrium conditions are then clearly the balancing Equations (5.34) which we repeat here for convenience:

$$D_j = kW_j \qquad (5.48)$$

We noted that D_j could be calculated from the model, using Equations (5.1), (5.2) and (5.4) and we derived Equation (5.36) in that way. Thus, the equilibrium equations to be solved for W_j, if we substitute from (5.36) into (5.48) are

$$\sum_i \frac{e_i P_i W_j^\alpha e^{-\beta c_{ij}}}{\sum_k W_k^\alpha e^{-\beta c_{ik}}} = kW_j \qquad (5.49)$$

These are clearly highly non-linear in the W_j's and cannot be solved analytically. However, it turns out that we can use some tricks to gain some insights into the nature of the solutions and their bifurcation properties.

The two sides of Equation (5.49) represent different ways of computing the revenue in centre j: the left hand side is obtained from the model of consumers' behaviour, the spatial interaction model; the right hand side represents suppliers' behaviour. If we define separate functions for each, calling them $D_j^{(1)}$ and $D_j^{(2)}$ respectively, then we have

118

$$D_j^{(1)} = \sum_i \frac{e_i P_i W_j^\alpha e^{-\beta c_{ij}}}{\sum_k W_k^\alpha e^{-\beta c_{ik}}} \qquad (5.50)$$

and

$$D_j^{(2)} = kW_j \qquad (5.51)$$

These can also be usefully called the 'revenue' and 'cost' curves respectively. If we focus on a single zone j, then we can plot $D_j^{(1)}$ and $D_j^{(2)}$ separately against W_j and note that the equilibrium points are the intersections of the two curves. This is simple for Equation (5.51): $D_j^{(2)}$ is a straight line if k is assumed to be constant. So our main initial task is to investigate $D_j^{(1)}$ as a function of W_j. The first complexity to note is that the right hand side of (5.50) is a function also of all other W_k's, $k \neq j$. However, we initially assume that these remain constant while W_j notionally varies. Later, we return to this question and investigate more realistic assumptions.

The initial step in the argument is to look at the first and second derivatives of $D_j^{(1)}$ with respect to W_j. Differentiation of (5.4), repeated once, gives

$$\frac{\partial D_j}{\partial W_j} = \sum_i \frac{\partial S_{ij}}{\partial W_j} \qquad (5.52)$$

and

$$\frac{\partial^2 D_j}{\partial W_j^2} = \sum_i \frac{\partial^2 S_{ij}}{\partial W_j^2} \qquad (5.53)$$

and this indicates that we can get the information we want once we know the derivatives of S_{ij} with respect to W_j. The details of the calculations are given in Appendix 3. Here, we just note the main results:

$$\frac{\partial S_{ij}}{\partial W_j} = \frac{\alpha S_{ij}}{W_j} \left[1 - \frac{S_{ij}}{e_i P_i} \right] \qquad (5.54)$$

119

and

$$\frac{\partial^2 S_{ij}}{\partial W_j^2} = \frac{\alpha S_{ij}}{W_j} \left[(\alpha - 1) + (1 - 3\alpha) \frac{S_{ij}}{e_i P_i} \right.$$
$$\left. + 2\alpha \left(\frac{S_{ij}}{e_i P_i} \right)^2 \right] \tag{5.55}$$

For the second derivative, the term in square brackets can be factorised to give

$$\frac{\partial^2 S_{ij}}{\partial W_j^2} = \frac{2\alpha^2 S_{ij}}{e_i P_i} \left(\frac{S_{ij}}{e_i P_i} - 1 \right) \left[\frac{S_{ij}}{e_i P_i} - \left(\frac{\alpha - 1}{2\alpha} \right) \right] \tag{5.56}$$

We can now examine the behaviour of these derivatives for a range of values of W_j. To do this, it is helpful to keep the basic model equation for S_{ij} in mind and the most convenient form for it involves combining (5.1) and (5.2) by substituting for A_i:

$$S_{ij} = \frac{e_i P_i W_j^\alpha e^{-\beta c_{ij}}}{\sum_k W_k^\alpha e^{-\beta c_{ik}}} \tag{5.57}$$

We can then note the following features of the first derivative, using Equations (5.54) and (5.57).

(1) As $W_j \to \infty$, (5.57) shows that the term $W_j^\alpha e^{-\beta c_{ij}}$ in the sum in the denominator dominates and is equal to the same term in the numerator. So, $S_{ij} \to e_i P_i$ and the term in square brackets in (5.54) then gives us our main result:

$$\frac{\partial S_{ij}}{\partial W_j} \to 0 \quad \text{as} \quad W_j \to \infty \tag{5.58}$$

(2) As $W_j \to 0$, $S_{ij} \to 0$ and hence so also does $S_{ij}/e_i P_i$. S_{ij} in Equation (5.54) has a factor W_j^α and so $\frac{\partial S_{ij}}{\partial W_j}$ has a factor $W_j^{\alpha-1}$. Thus:

$$\frac{\partial S_{ij}}{\partial W_j} = \begin{cases} 0 & \text{if } \alpha > 1 \\ \text{finite (and} & \text{if } \alpha = 1 \\ \quad \text{positive)} \\ \infty & \text{if } \alpha < 1 \end{cases} \quad (5.59)$$

(3) For values of W_j between zero and infinity, we simply note from (5.54) that $\frac{\partial S_{ij}}{\partial W_j}$ is always positive (since $S_{ij}/e_i P_i < 1$).

The equivalent results for the second derivative are as follows:

(1) As $W_j \to 0$, $S_{ij} \to 0$ and so Equation (5.56) shows that $\frac{\partial^2 S_{ij}}{\partial W_j^2} \to 0$. It can also be seen from this equation that for small non-zero values of W_j, $\frac{\partial^2 S_{ij}}{\partial W_j^2}$ is less than zero if $\alpha < 1$ and greater than zero if $\alpha > 1$.

(2) As $W_j \to \infty$, then $S_{ij} \to e_i P_i$ and so again $\frac{\partial^2 S_{ij}}{\partial W_j^2} \to 0$.

(3) For finite values of W_j (greater than zero of course), we note from Equation (5.56) that the first factor is positive, the first factor in brackets is negative and therefore the overall sign depends on the second factor in brackets. This involves us in a study of the term $\frac{\alpha - 1}{2\alpha}$ which appears in that factor. This is plotted in Figure 5.5. This shows clearly that for $\alpha < 1$, the second factor is always positive and hence $\frac{\partial^2 S_{ij}}{\partial W_j^2}$ is negative for all W_j.

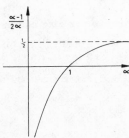

Figure 5.5 Plot of $(\alpha-1)/2\alpha$

For $\alpha > 1$, the term $\frac{\alpha - 1}{2\alpha}$ is positive and so the sign of the factor depends on S_{ij}/e_iP_i relative to it. If we put

$$x = \frac{\alpha - 1}{2\alpha} \tag{5.60}$$

then the appropriate results can be collected together as follows:

For $\alpha > 1$: $\dfrac{\partial^2 S_{ij}}{\partial W_j^2}$ = $\begin{cases} \text{positive if } \dfrac{S_{ij}}{e_iP_i} < x \\[2ex] 0 \qquad\quad \text{if } \dfrac{S_{ij}}{e_iP_i} = x \\[2ex] \text{negative if } \dfrac{S_{ij}}{e_iP_i} > x \end{cases}$ (5.61)

Thus, for $\alpha < 1$, S_{ij} has no points of inflexion when plotted against W_j; for $\alpha > 1$, it has one point of inflexion when $\dfrac{S_{ij}}{e_iP_i} = \dfrac{\alpha - 1}{2\alpha}$

This information now allows us to plot the shape of the $S_{ij} - W_j$ curve for different values of α and this is done in Figure 5.6.

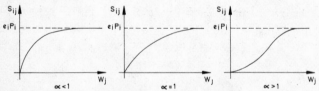

Figure 5.6 Retailing : flow-centre size curves

We can now examine the plots of D_j against W_j using Equations (5.52) and (5.53) and substituting the information we have now acquired about the first and second derivatives of S_{ij}, and the shape of the $S_{ij} - W_j$ curves. The expressions for the derivatives can be written:

$$\frac{\partial D_1}{\partial W_j} = \sum_i \frac{\alpha S_{ij}}{W_j} \left(1 - \frac{S_{ij}}{e_i P_i}\right) \tag{5.62}$$

$$= \frac{\alpha}{W_j} D_j \left(1 - \frac{1}{D_j} \sum_i \frac{S_{ij}^2}{e_i P_i}\right) \tag{5.63}$$

and

$$\frac{\partial^2 D_j}{\partial W_j} = (\alpha - 1) \sum_i \frac{\alpha S_{ij}}{W_j} + (1 - 3\alpha) \sum_i \frac{\alpha S_{ij}^2}{W_i e_i P_i}$$

$$+ 2\alpha \sum_i \frac{\alpha S_{ij}^2}{W_j e_i^2 P_i^2} \tag{5.64}$$

It is straightforward to show that

$$\text{as} \quad W_j \to 0, \quad \frac{\partial D_j}{\partial W_j} \to \begin{cases} \infty, & \alpha < 1 \\ \text{finite (positive)}, & \alpha = 1 \\ 0, & \alpha > 1 \end{cases} \tag{5.65}$$

and

$$\frac{\partial D_j}{\partial W_j} \to 0 \quad \text{as} \quad W_j \to \infty \tag{5.66}$$

The argument in relation to the second derivative is more complicated, because it can no longer be factored. But since the D_j - W_j curve is obviously a sum of S_{ij} - W_j curves, we can deduce that they have broadly the same shapes, but we must note a difficulty: the points of inflexion in the S_{ij} - W_j curves can occur at different places for each i. Indeed, we can find, for each i, the value of W_j corresponding to the inflexion - call it $W_j^{i \, \text{inflex}}$, with a notation showing the dependence on i. It occurs when

$$S_{ij} = x e_i P_i, \quad \text{where} \quad x = \frac{\alpha - 1}{2\alpha} \tag{5.67}$$

which, writing S_{ij} in full from (5.57) can be given as

$$x e_i P_i = \frac{e_i P_i W_j^\alpha e^{-\beta c_{ij}}}{\sum_k W_k^\alpha e^{-\beta c_{ik}}} \tag{5.68}$$

Chapter 5

After some manipulation, this gives

$$W_j^i \text{ inflex} = \left[\frac{x \sum\limits_{k \neq j} W_j^\alpha e^{-\beta c_{ik}}}{(1 - x)e^{-\beta c_{ij}}} \right]^{1/\alpha} \tag{5.69}$$

The S_{ij} - W_j curves, with the points of inflexion in different places, are shown in Figure 5.7. This suggests that, when they are added together, there may be more than one point of inflexion in the D_j - W_j curve. For $\alpha < 1$, $\frac{\partial^2 D_j}{\partial W_j^2} < 0$ since it is made up of all such negative terms.

Thus, we can now collect these results together and present the forms of the D_j - W_j curves for different α values in Figure 5.8. Note that the upper bound is now, of course $\sum\limits_i e_i P_i$.

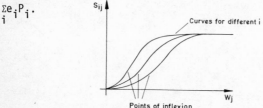

Figure 5.7 Points of inflection on flow-size curves

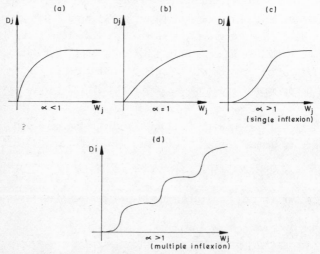

Figure 5.8 Retailing : revenue-size curves

124

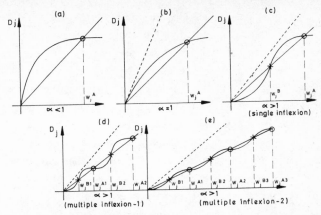

Figure 5.9 Revenue-size curves

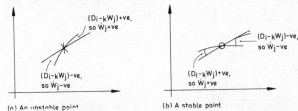

Figure 5.10 Stability considerations - revenue-cost-size curves

It is now possible to move to the heart of the argument: by adding the lines $D_j = kW_j$ to the curves of Figure 5.8, we can examine the equilibrium points which are the intersections of the $D_j^{(1)}$ curve and the $D_j^{(2)}$ line for each case and investigate their stability. The basis for this is Figure 5.9. The two kinds of intersection of curve and line are shown in Figure 5.10. When the curve is above the line, $D_j - kW_j$ is positive, and hence so is $\dot{W}_j$; and vice versa. This means that in case (a), W_j increases above the intersection and decreases below it, and hence the point is unstable. The reverse argument applies in case (b), which is a stable point. This argument is applied to the various intersections on Figure 5.9: the stable points are circled and the unstable points marked with crosses.

The next point to note is that the line does not always intersect the curve: two cases are shown on each of the plots

125

in Figure 5.9 - the dashed line obviously being the non-intersecting case. An additional example has been added to Figure 5.9 - case (e) - which shows an intermediate case of the line intersecting part of the curve, but not all of its as in case (d). This will be the basis of our initial discussion of bifurcation properties. First, however, we concentrate on the intersecting case and examine stability.

The origin, $W_j = 0$, is unstable in the cases where $\alpha < 1$ and is stable in the case $\alpha > 1$. Otherwise, the points marked W_j^A are stable and those marked W_j^B are unstable. (In the multiple-inflexion cases, these are marked W_j^{A1}, W_j^{A2}, ..., and W_j^{B1}, W_j^{B2}, ..., respectively). The reasons for this should be clear from the argument associated with Figure 5.10. One point to emphasise is that the argument leads us to expect *no* stable equilibrium point at the origin for the case $\alpha < 1$.

To illustrate the idea of bifurcation, we first focus on the $\alpha > 1$ case, and on changing values of the parameter k. Recall that this parameter represents something like the cost of supplying a unit of shopping centre space, and that this might change over time, as a result of technological develop-ments or new rent patterns say. The $D_j = kW_j$ line is plotted for three different k values in Figure 5.11. In case (1), there are two possible stable points; in case (3), there is only one stable point, the origin; and case (2), the value of k is critical. The situation changes from one where develop-ment is possible in the zone (for lower values of k) to one where it is not (for higher values). The critical value of k obviously depends on the zone j and so can be labelled k_j^{crit}.

The next stage in the argument is to note that k appears to play a special role among the parameters because it repre-sents the slope of the line which figures prominently in the geometrical interpretation. However, it is not really so special: when the zone is in a critical state, with the cost

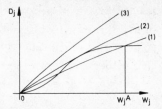

Figure 5.11 Revenue curve with varying cost lines

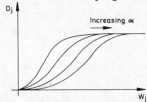

Figure 5.12 Effect of increasing α on revenue curve

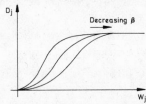

Figure 5.13 Effect of decreasing β on revenue curve

line touching the revenue curve, then the situation is critical
with respect to all parameters.   A change in any of them,
other than k, would shift the position of the revenue curve and
shift it away from the critical position. For example, we show
in Figure 5.12 what happens to the curve as α increases and in
Figure 5.13 what happens as β decreases.   Broadly speaking, as
α increases, the curve steepens for higher values of W_j as
shown.   This means that k_j^{crit} it lower in such cases which
means that, other things being equal, there is less likely to
be a centre in a particular zone (because the k-parameter for
the whole system has a greater probability of exceeding k_j^{crit}
in particular zones).   The argument for β is the opposite; as
β increases, consumers are less likely to travel far; so k_j^{crit}

increases with β and this makes development more likely in any zone. In other words, this provides an interpretation of the well-known result that higher α and lower β combine to produce patterns with smaller numbers of larger centres.

This analysis shows that there is a set of points, which we can label $(k_j^{crit}, \alpha_j^{crit}, \beta_j^{crit})$ in the parameter space at which the zone is in a critical state. This set forms a surface. To get some idea of its shape, consider first the relation between k_j^{crit} and β_j^{crit} for fixed α. As β_j^{crit} increases, we can see from Figure 5.13 that k_j^{crit} also increases, probably from a minimum to infinity, since when β → ∞, all trips will go to the nearest centre, in which case the tangential critical line will be vertical. This is shown in Figure 5.14. We can expect a similar relationship between k_j^{crit} and α_j^{crit}, but in the reverse direction. This is sketched in Figure 5.15. Since, however, it is more difficult to sketch the relationship between α_j^{crit} and β_j^{crit} for fixed k, this means that we do not as yet have complete information to sketch the whole surface.

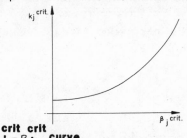

Figure 5.14 $k_j^{crit} - \beta_j^{crit}$ curve

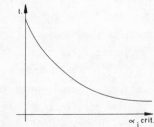

Figure 5.15 $k_j^{crit} - \alpha_j^{crit}$ curve

The results on criticality can be expressed in another way which links with catastrophe theory. First, let us return to the situation where we are considering the variation of possible equilibrium values with k in the $\alpha > 1$ case. In Figure 5.11 the point W_j^A, when it exists, and the origin, are the stable equilibrium points. We can plot the values of these against a varying k value. This is done in Figure 5.16. This curve is immediately reminiscent of the fold catastrophe, but with zero states added. We could also then do corresponding plots for α and β and this is done in Figures 5.17 and 5.18. In each case, they assume that k and the other parameters are fixed. The

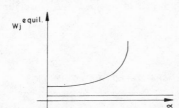

Figure 5.16 Retailing centre size and the fold catastrophe, k as control variable

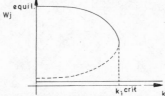

Figure 5.17 Retailing centre size and the fold catastrophe, α as control variable

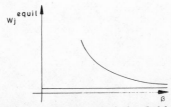

Figure 5.18 Retailing centre size and the fold catastrophe, β as control variable

basis of these plots can be worked out by adding a k-line in Figures 5.12 and 5.13 in turn and examining the variation in W_j^A against the corresponding parameter.

The notion of simultaneous criticality with respect to all parameters can also be expressed more formally in a mathematical sense. If we add W_j-variation to the Lagrangian given in Equation (A1.11) of Appendix 1 we can see the equilibrium values of W_j as arising from

$$\underset{\{S_{ij},W_j\}}{\text{Max}} \quad L = \log S! - \underset{ij}{\Sigma} \log S_{ij}! + \underset{i}{\Sigma}\lambda_i(e_iP_i - \underset{j}{\Sigma}S_{ij})$$

$$+ \alpha(\underset{ij}{\Sigma}S_{ij} \log W_j - \overline{\log W})$$

$$+ \beta(C - \underset{ij}{\Sigma}T_{ij}c_{ij}) + \gamma(W - \underset{j}{\Sigma}W_j) \qquad (5.70)$$

The Lagrangian multiplier γ has a 1-1 relationship with our parameter k. So, formally, we can write (5.70) as

$$\underset{\{S_{ij},W_j\}}{\text{Max}} \quad L = L(W_j,k,\alpha,\beta, \text{ other parameters taken as fixed}) \qquad (5.71)$$

where we now use k instead of γ. The equilibrium solution is

$$\frac{\partial L}{\partial W_j} = 0 \qquad (5.72)$$

which can be written more conveniently as

$$f_j(W_j,k,\alpha,\beta, \dots) = 0 \qquad (5.73)$$

for a suitably defined set of functions, f_j. An equilibrium point disappears when

$$\frac{\partial^2 f}{\partial W_j^2} = 0 \qquad (5.74)$$

which means that

$$\frac{\partial f_j}{\partial W_j} = 0 \qquad (5.75)$$

at what we have called critical points. (We concentrate on these first order critical points below, but we should also bear in mind the possibility of higher order critical points, occurring for example, when the following determinants vanish:

$$\left[\frac{\partial f_j}{\partial W_k}\right] = \left[\frac{\partial^2 f}{\partial W_j \partial W_k}\right] = 0.$$

We can show Equations (5.75) explicitly and give it a geometrical interpretation for k-criticality. Let us differentiate L in (5.70) with respect to W_j to get one of the equilibrium conditions:

$$\frac{\alpha}{W_j} \sum_i S_{ij} - \gamma = 0 \qquad (5.76)$$

which can be rewritten as

$$f_j = D_j - kW_j = 0 \qquad (5.77)$$

which shows that

$$k = \frac{\gamma}{\alpha} \qquad (5.78)$$

Recall also that

$$D_j = \sum_i S_{ij} \qquad (5.79)$$

We can now write the condition for criticality, (5.75) explicitly using (5.77) as

$$\frac{\partial D_j}{\partial W_j} - k = 0 \qquad (5.80)$$

and this is simply stating that the slope of the revenue curve is equal to the slope of the line when the zone is in the critical state.

A theorem on the derivatives of implicit functions (*cf.* for example, Yamane, 1968, p. 154) shows that[2]

$$\frac{\partial k}{\partial W_j} = -\frac{\partial f_j}{\partial W_j} \Big/ \frac{\partial f_j}{\partial k} \qquad (5.81)$$

Chapter 5

and so, because of (5.75), we must have

$$\frac{\partial k}{\partial W_j} = 0 \tag{5.82}$$

This is simply an algebraic statement that when the equilibrium value of W_j, which we can label W_j^{equil}, is plotted against k, as in Figure 5.16, the tangent to the curve is vertical at the critical points.

More importantly, this is true for any parameter. Let z be an arbitrary parameter of f_j. Then

$$\frac{\partial z}{\partial W_j} = - \frac{\partial f_j}{\partial W_j} \Bigg/ \frac{\partial f_j}{\partial z} \tag{5.83}$$

We have already seen that this applies to α and β and we have discussed the consequences of this above. But it also applies to any other variables we have considered fixed - for example to e_i, P_i or c_{ij}: any disturbance in one of these at a critical point and there is a bifurcation. The next step in the argument, therefore, is to consider the effect of e_i or P_i variation.

We begin with offering a geometrical interpretation of the effects of e_i or P_i changing. In practice, it is easier to consider them together, as $e_i P_i$, since a change in either or both of them will operate in the same way. We can use an argument analogous to that which employed the definition of the functions $D_j^{(1)}$ and $D_j^{(2)}$ and seek equilibrium points as an intersection of the two curves. Let us focus on the effect of one term, $e_I P_I$, in zone I, on zone j. The equilibrium condition can be written as

$$D_j - kW_j = 0 = \sum_i S_{ij} - kW_j$$

$$= S_{Ij} - (kW_j - \sum_{i \neq I} S_{ij}) \tag{5.84}$$

where, now, only the S_{Ij} term contains $e_I P_I$. Thus the equilibrium condition can be written as

$$F_{Ij}^{(1)} = F_{Ij}^{(2)} \tag{5.85}$$

where

$$F_{Ij}^{(1)} = S_{Ij} \tag{5.86}$$

and

$$F_{Ij}^{(2)} = kW_j - \sum_{i \neq I} S_{ij} \tag{5.87}$$

$$= kW_j - G_j(W_j, k, \alpha, \beta, \ldots) \tag{5.88}$$

for a suitable definition of the functions G_j. We have already seen that any S_{ij} term exhibits logistic-type behaviour against W_j variation, and so this gives us the form of $F_{Ij}^{(1)}$ as plotted in Figure 5.19. The function G_j is a sum of such functions and we will assume for the present argument that it has a single point of inflexion (though we should obviously consider how to modify the argument in the less likely event of it having several). This is plotted in Figure 5.20 along with the kW_j line. This figure also shows the result of subtracting these and the result is a plot of $F_{Ij}^{(2)}$. Two obviously different cases are shown. These, together with the critical case, are plotted with the $F_{Ij}^{(1)}$ curve in Figure 5.21. The two curves touch when the zone is in a critical state. And here, of course, we have

$$\frac{\partial F_{Ij}^{(1)}}{\partial W_j} = \frac{\partial F_{Ij}^{(1)}}{\partial W_j} \tag{5.89}$$

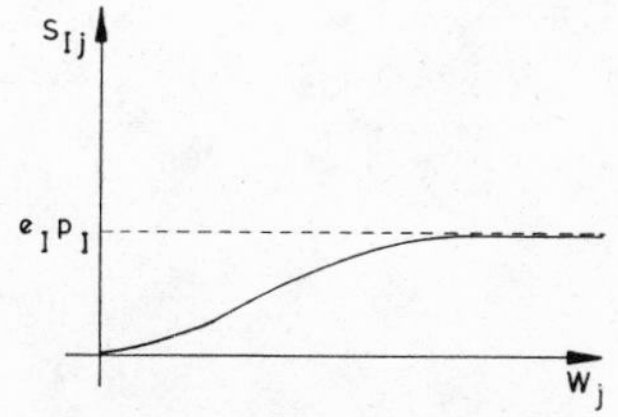

Figure 5.19 Flow vs. centre size

Chapter 5

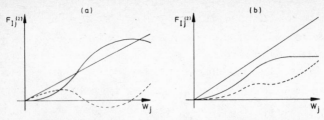

Figure 5.20 Plot of revenue curves, less one flow, relative to cost

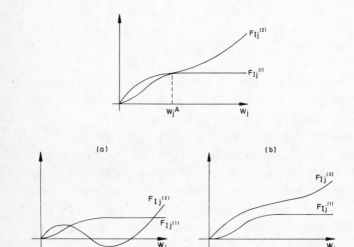

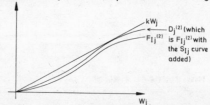

Figure 5.21 Flow, revenue less one flow: two cases and the critical case

which is simply another way of writing the equilibrium condition. (Note that kW_j line and the $F_{Ij}^{(2)}$ curve do not touch at the critical point because kW_j touches $D_j^{(1)}$ which is similar to $F_{Ij}^{(2)}$ but which also has the S_{Ij} contribution. This is shown diagrammatically, for completeness in Figure 5.22.

Figure 5.22 Critical case : alternative presentation

We can now use these diagrams to get some insight into the effect of $e_I P_I$ changes. Since this term is a simple multiplicative factor in S_{Ij}, if $e_I P_I$ increases, the curve is pushed up, and vice versa. Thus, in Figure 5.21(b), increasing $e_I P_I$ could push up the $F_{Ij}^{(1)}$ curve until it touched the $F_{Ij}^{(2)}$ curve after which development at j would become possible.

Two steps remain in this particular argument: the c_{ij} variables and the W_k, $k \neq j$, variables. The first is much simpler and so we get it out of the way. c_{ij} only appears in the cost deterrence term, $e^{-\beta c_{ij}}$ and so when it changes, it is mainly as though a factor in S_{ij} changes. We should also bear in mind that it appears in the donominator, though as one term among many and so a particular change is not likely to have any great impact there. In the main, therefore, the effect of a c_{ij} change will be rather like that of an $e_i P_i$ change. There is one possible exception to this: if there is a general change in the 'ease of travel', then this is much more like a β change, because it scales all the c_{ij}'s. In that case, therefore, c_{ij} changes are better treated as quasi-β changes.

W_k, $k \neq j$, changes are very much more difficult to handle. So far, we have focused on a single zone, j, but it is clear that if other W_k's change, then they could have a major impact on the revenue curve. Indeed, a fundamental problem should be recognised at the outset: to carry out the above analysis at all involves the assumption that all the W_k's are fixed and constant as W_j varies over a wide range. This is unlikely ever to happen in practice and we will see below that this leads to a modification of the methods of setting up the revenue curve to examine criticality. First, however, we examine the problem of deciding what it means to have W_k, $k \neq j$, 'fixed'. The argument below largely follows that of Wilson and Clarke (1979).

We can usefully think of all the other W_k's as providing a 'backcloth' of competition for any particular zone j. To illustrate that the problem of fixing the backcloth is a relatively difficult one, we explore separately a number of ideas as to what it could mean.

135

Chapter 5

(1) Consider first the assumption that the W_k, $k \neq j$, are fixed. The way the calculation of the revenue D_j - W_j curve then proceeds is as follows. An initial set of values are assumed for W_k, $k \neq j$, which we can call W_k^{init}. These can be fed into the spatial interaction model (5.1) for each of a range of W_j values. D_j can be calculated from the resulting S_{ij}'s. This does indeed turn out to generate curves of the predicted form for the different α-values (see Wilson and Clarke, 1979). However, there are two basic problems with this procedure, one general but not disabling, the other demanding a modification to the calculation procedures associated with this assumption.

The general problem is that

$$W = \sum_j W_j \qquad (5.90)$$

does not remain constant. Note that

$$\sum_j D_j = \sum_i e_i P_i = k \sum_j W_j \qquad (5.91)$$

and hence

$$k = \frac{\sum_i e_i P_i}{\sum_j W_j} \qquad (5.92)$$

The sum, $\sum_j W_j$ is made up of all the fixed W_k^{init}'s together with W_j and therefore increases as W_j increases. This means that k slowly decreases. Then, instead of $D_j = kW_j$ being a straight line, it bends slightly as shown in Figure 5.23. However it can still touch the logistic D_j - W_j curve at criticality in

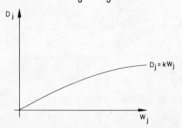

Figure 5 .23 Cost line transformed to cost curve

exactly the same way as a straight line and so this modification does not change the general form of the argument.

The second problem is this: with the calculation procedure used, if the W_k^{init}'s are selected arbitrarily, when there is a non-zero equilibrium point for W_j, say W_j^A , then while

$$D_j = kW_j^A \qquad\qquad (5.93)$$

at that point

$$D_k \neq kW_k^{init} \qquad\qquad (5.94)$$

because of the arbitrary selection of these values. In other words, the point $(W^{init}, ..., W_j^A, ..., W_N^{init})$ is not, typically, a system equilibrium point. The task now is to find a set of W_k^{init}'s so that when W_j^A is found, the whole set of W_k's do constitute such a point. The resulting modified assumption is what was referred to in Wilson and Clarke (1979) as Assumption 1'. It works as follows. Find a critical point by running a suitable algorithm for calculating equilibrium points and noting some parameter values at which a jump takes place. Take the W_k, $k \neq j$, which occur just before or after the point (and unfortunately, there is no reason why they should be the same) and use these to build the $D_j - W_j$ curve using the method first outlined above. In this way, it should be possible to construct a curve, and the modified version of the $D_j = kW_j$ 'line' (itself now a curve as a result of Equation (5.92) and the assumption) at which the critical equilibrium point is generated.

The inherent difficulty of this method is that it is necessary to locate a 'jump', or critical point, before it is possible to apply it, and this takes away some of the point of the exercise. It allows us to interpret that particular jump, but does not provide a method for discovering it in the first place.

(2) A more general procedure might arise out of attempting to answer the question of how to generate the $D_j - W_j$ curve around an equilibrium point which is not critical. If such a point is generated, say from a mathematical programme, then

this can be used to supply a set of W_k^{init}'s to calculate other
points on the D_j - W_j revenue curve. Such points are only
likely to be valid in the neighbourhood of the equilibrium
point itself since over a wider range, as noted earlier, there
is no reason why the W_k's $k \neq j$, are likely to remain fixed.
However it may suffice to examine only a small section of the
curve as depicted in Figure 5.24, first for the critical case,
and then for the more general case now being considered. This
may then provide the basis for finding critical points. For
example, the angle between the two curves may provide a useful
measure as to how near the zone is to a critical condition, and
this is also indicated on the figure.

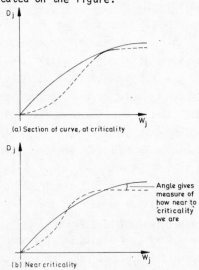

(a) Section of curve, at criticality

(b) Near criticality

Angle gives
measure of
how near to
criticality
we are

Figure 5.24 Criticality and near-criticality with cost curve

One other difficulty can usefully be mentioned at this
point: in our earlier argument, we noted that the revenue
curve and the cost curve do not always intersect. In this
case, there is no equilibrium solution with a non-zero W_j.
This leads to even greater difficulties as to what should be
used for a backcloth: perhaps a set of W_k's generated with the

additional constraint that $W_j = 0$. However, this question is relatively unimportant as W_j is known to be zero in such cases: only the actual critical case and the intersecting cases demand a detailed knowledge of the curves.

(3) Another method is to take a set of W_k^{init}'s from some real data: there is a good chance that these would then be near their equilibrium values and that the computation of the $D_j^{(1)}$ and $D_j^{(2)}$ curves might then intersect in a genuine overall equilibrium point.

These appear to be the only reasonable ways of proceeding known at the present time. Two other assumptions were considered in the Wilson and Clarke (1979) paper but they are unrealistic in a number of respects and are referred to only briefly here:

(4) For each W_j, we could fix this and then take the set of W_k's, $k \neq j$, as being determined by the mathematical programme. The problem with this assumption is that the backcloth is then different for each point of the revenue curve.

(5) The other alternative is to fix total W and then to let all the W_k's, $k \neq j$, decrease proportionately as W_j increases. The problem with this assumption is that as $W_j \rightarrow W$, all the other W_k's tend to zero, and this leads to highly unrealistic revenue curves for high values of W_j.

Two further points need to be made to round off the argument. First, in investigating bifurcation, we have concentrated on the $\alpha > 1$ case. This obscures the very important result that $\alpha = 1$ is clearly a critical point in parameter space. When $\alpha < 1$, all transitions are smooth as a result of smooth parameter changes; when $\alpha > 1$, jumps are possible. This makes α reminiscent of the splitting factor in the cusp catastrophe, described in Chapter 1. However, we have also seen, that, as a result of shifts in the revenue curve, changes in α can also cause 'normal' effects. α, therefore, cannot be a splitting factor in a canonical sense. This does illustrate, though, that we have concentrated on 'first order' criticality.

Chapter 5

This is probably the most important feature from a practical point of view, but we noted earlier that higher order critical points in parameter space may exist. The above argument about α suggests that these may be at least cusp points (though it has been argued by Amson, 1979, that this point looks like a pucker point and not a Thom-like cusp - a matter which awaits full clarification). If we add β to the list of parameters, then Thomian arguments would take us in the direction of possible swallowtail catastrophes. But we have also seen that there are, in effect, many parameters, and this takes us outside the realms of elementary catastrophe theory into generalised catastrophe theory which is not very well understood. So the first general point is to recognise that $\alpha = 1$ is critical and represents some sort of cusp, and this, together with general arguments, warns us that there may be even higher order singularities around.

The second general point is to emphasise the nature of the bifurcation properties which have been established in one respect. In all cases where we have established that a non-zero value of W_j is possible, it has also remained true that the zero state remains a possibility. We might describe one state, for a zone, as the 'development possible' - or DP - state and the other the 'no development possible' - or NDP - state. The point to emphasise here is that when development is possible, W_j is not necessarily non-zero. Since this kind of analysis applies to each zone in turn, the general deduction to be made from this analysis is that there are likely to be many possible equilibrium states for the whole system associated with any possible set of parameter values, and these are distinguished by the zeros and their locations.

There will be the greatest number of alternatives when the parameters are such that all zones are in the DP state: any number or combination of zeros is then also a possible pattern. We might also conjecture that this state holds when the parameters are within some (closed?) region of parameter space.

Outside this space, one or more zones have to have zero W_j: they are in the NDP state, and this reduces the number of combinations relative to the system state with the maximum number of non-zero elements. All this means that, when modelling development, a rule has to be supplied to say which of the different possible equilibrium states is adopted. This may be to do with the maximisation of consumers' surplus. It may postulate that the maximum possible number of zones have non-zero W_j's (though this may not be in itself sufficient to pick out a unique system state).

It does seem likely that when the state with all non-zero W_j's exists, then it is unique, but this is an issue which we postpone to Section 5.4.7 below, when we discuss ecological analogies and seek to become equipped to explore the solutions of the equations in a more topological sense.

5.4.3 Summary of results on retailing : towards a theory of structural evolution

The argument of Section 5.4.2 has been long and complicated. It may be useful therefore to summarise the conclusions and to see what contribution they begin to make towards a theory of the evolution of urban structure.

First, our analyses confirm the results of other authors (such as Harris, 1965; and White, 1977, 1978) on the kinds of centre structures which result from various types of parameter values. For example, high α and low β encourages a system with a relatively small number of large centres, and vice versa. We can also add a note on the effect of other parameters in this respect. For example, high k values will also tend to reduce the number of centres. Similarly higher e_i and P_i values will obviously affect the growth of the system as a whole and may affect the number of centres by allowing thresholds to be crossed. The main general result to emerge is that, for a particular zone at a particular time, there is a surface in parameter space on one 'side' of which development is possible in that zone, on the other side not. These were called DP and NDP states for that zone. This was seen most clearly with the role of the k-parameter, but the argument was extended to include every parameter and variable which was exogenous to the main problem. We saw also that $\alpha = 1$ was an important critical value: for $\alpha < 1$, zones are always in the DP state: for $\alpha > 1$ this is not the case.

When a zone switches from the NDP to the DP state, if development actually takes place, this will be recorded as a 'jump' in that particular W_j value. There will also be complications for analysis (and observation) arising from the fact that such a jump will also cause secondary jumps in the other W_k variables. We also saw that there is at least one other possible cause of jumps: if the attractiveness factor takes a particular form, there is the Poston and Wilson (1977) mechanism which can generate jumps as a result of β changes. It turns out that there are other causes of jumps also. Because the equilibrium problem can be formulated as a mathematical programming problem, these may result from what Poston and Stewart (1978) call 'constraint catastrophes'. We will also see when we pursue the ecological analogy in Section 5.4.7 that, following a disturbance from equilibrium, there may be separatrix-crossing jumps. This means that the interpretation of jumps, both in theory and practice, is a difficult exercise, and we will return to this issue later.

A particularly difficult task, in carrying out the analysis for a particular zone, was to decide on the nature of the W_k, $k \neq j$, backcloth for the zone. This is at the heart of the task of describing, through these mechanisms, the evolution of the structure of the whole system: in describing what happens in a particular zone, we have to build in what is happening in all other zones. Another important point can usefully be noted at this stage: any ad hoc development - say the building of a large shopping centre by some particular entrepreneur - can obviously have a major effect on the development of the system as a whole. In this sense, the predictive modeller is faced with the task of building in 'historical accidents'. This also demonstrates the truism that there are many possible particular paths of evolution.

The task of modelling development, therefore, can perhaps be reformulated as follows. Any particular path of evolution can be interpreted using the kinds of models outlined above, and this will include the building in of various ad hoc decisions. There may be some regularities in the patterns and structures at various times which can be predicted in a deterministic way. These would include results on the overall size distributions of centres for example. This in itself will be important from a planning viewpoint. The planner may seek to achieve combinations of parameters which ensure that an overall form of structure develops even though it is not possible to guarantee the particular spatial form in detail. Also, it is likely to be useful to the planner to exploit the DP-NDP concept for particular zones, and to take action which will encourage or discourage, as required, development in particular places by manipulating parameters.

What would an account of the evolution of retailing structure then look like, using the models set up in the previous subsection? From any particular initial state, it would proceed in relatively short time intervals. Between each one, any exogenous variables would be updated and it would be possible to calculate the DP-NDP states for particular zones. Directions of change could be calculated and noted: this would include any jumps and also whether particular zones are approaching criticality (and this connects to the notion of analysing robustness and resilience of parts of the system).

The particular methods used would depend on whether a historical sequence was being analysed or whether the model was being run predictively. In the former case, it would be a matter of examining the extent to which the system's development conformed with the broad predictions of the model, possibly modified to include ad hoc events - 'historical accidents'. In the latter case, a rule would have to be supplied for choosing

143

the particular equilibrium state from among the many possible for each step forward. This should then be interpreted not as a detailed spatial prediction, but as an indication of the overall form of structure which is likely to develop.

5.4.4 Residential location

Although residential location is in principle more complicated than the retail case, we treat it rather more briefly, partly because the most elementary case can be treated directly by analogy, but then because the more realistic models are as yet too complicated to allow analysis to proceed very far.

The model which is analogous to the retail model was presented earlier as Equation (5.46). The equilibrium conditions is clearly

$$P_i = qH_i \tag{5.95}$$

which can be elaborated as

$$\sum_j \frac{B_j H_i E_j e^{-\mu c_{ij}}}{\sum_k H_i e^{-\mu c_{ik}}} = qH_i \tag{5.96}$$

and it can be seen that the argument could then proceed exactly in an analogous fashion to the one of Section 5.4.2 with $\alpha = 1$. It is more interesting to see what happens when the models are modified first with the introduction of composite attractiveness factors and secondly to couple this with some disaggregation.

An aggregate model with a composite attractiveness factor would take the form

$$T_{ij} = B_j W_i^{res} E_j e^{-\mu c_{ij}} \tag{5.97}$$

with

$$W_i^{res} = X_{1i}^{\alpha 1} X_{2i}^{\alpha 2} X_{3i}^{\alpha 3} \ldots \tag{5.98}$$

For example, we might take

$$X_{1i} = H_i \tag{5.99}$$

$$X_{2i} = L_i^{res} \tag{5.100}$$

$$X_{3i} = H_i / L_i^{res} \tag{5.101}$$

144

$$X_{4i} = H_i/L_i^{tot} \tag{5.102}$$

$$X_{5i} = (\sum_j W_j e^{-\beta c_{ij}}) \tag{5.103}$$

where L_i^{res} is the amount of land currently developed for housing, and L_i^{tot} is the total land area potentially available for housing. Then, X_{1i} measures the current availability of housing, X_{2i} the availability of land, X_{3i} the residential density of current development, X_{4i} the overall residential density and X_{5i} the accessibility to shops. These concepts are introduced in this form as an illustration, and much detailed work would have to be done to find a satisfactory empirical form for such attractiveness terms. But they do serve to illustrate what would have to happen in the dynamic model. The quantities which now change under development are H_i and L_i^{res} and we need a differential equation for each:

$$\dot{H}_i = \rho(\frac{h}{\lambda} \sum_j T_{ij} - qH_i) \tag{5.104}$$

and

$$\dot{L}_i^{res} = \sigma(\ell\sum_j T_{ij} - rL_i^{res}) \tag{5.105}$$

where ρ and σ are constants which measure the rate at which equilibrium is achieved; λ is the average number of workers per household and h is an average 'rent' per household; ℓ is the average rent component attributable to land; q and r are the costs of supplying new houses and units of land respectively. Again, in a realistic model, much work would have to be done on these constants.

The equilibrium conditions, written out in full using the various definitions can now be seen to be

$$\frac{h}{\lambda} \sum_j \frac{B_j E_j H_i^{(\alpha_1+\alpha_3+\alpha_4)} L_i^{res\,(\alpha_2-\alpha_3)} L_i^{tot-\alpha_4} (\sum_j W_j e^{-\beta c_{ij}})^{\alpha_5} e^{-\mu c_{ij}}}{\sum_k H_k^{(\alpha_1+\alpha_3+\alpha_4)} L_k^{res\,(\alpha_2-\alpha_3)} L_k^{tot-\alpha_4} (\sum_j W_j e^{-\beta c_{ij}})^{\alpha_5} e^{-\beta c_{ki}}} = qH_i \tag{5.106}$$

145

Chapter 5

$$\ell \sum_j \frac{B_j E_j H_i^{(\alpha_1+\alpha_3+\alpha_4)} L_i^{res(\alpha_2-\alpha_3)} L_i^{tot-\alpha_4} (\sum_j W_j e^{-\beta c_{ij}})^{\alpha_5} e^{-\mu c_{ij}}}{\sum_k H_k^{(\alpha_1+\alpha_2+\alpha_4)} L_k^{res(\alpha_2-\alpha_3)} L_k^{tot-\alpha_4} (\sum_j W_j e^{-\beta c_{ij}})^{\alpha_5} e^{-\beta c_{ki}}} = r L_i^{res}$$

(5.107)

In the first of these equations $(\alpha_1+\alpha_3+\alpha_4)$ plays the role of the α parameter in the retail case. The other factors act rather as in the manner of the $e_i P_i$ terms in the retail case. The situation with the second equation is similar with $\alpha_2-\alpha_3$ playing the role of the α factor. In both equations, the introduction of the shopping accessibility term is interesting in that it would couple this model to the retail model and any jumps in the W_j's would have secondary effects in this model. The other points of interest arise particularly in the second equation: if residential density was more important than land availability, then $\alpha_2-\alpha_3$ would be negative and this would take us into the realm of models with negative 'α' parameters.

The next stage in the argument is to look at disaggregation. This is particularly important here because it is evident that similar groups of people do in practice cluster together in houses of particular types. Rather than attempt a general analysis of this case, we examine a two person type, two house type model: rich and poor people, good and bad quality housing. Let the superscript k and w represent house type and person type respectively, but assume that only rich people live in good houses and only poor in bad houses. Then the models are

$$T_{ij}^{11} = B_j^1 E_j^1 W_i^{res1} e^{-\mu^1 c_{ij}}$$

(5.108)

and

$$T_{ij}^{22} = B_j^2 E_j^2 W_i^{res2} e^{-\mu^2 c_{ij}}$$

(5.109)

and for illustrative purposes we could take the attractiveness factors to be

$$W_i^{res1} = (H_i^1)^{\alpha_1^1} (1 + H_i^2)^{-\alpha_2^1}$$

(5.110)

and

$$W_i^{res2} = (H_i^1)^{\alpha_1^2} (H_i^2)^{\alpha_2^2}$$

(5.111)

146

The 1 is added in the $(1 + H_i^2)^{-\alpha_2^1}$ term to avoid problems when H_i^2 is zero. This means that the rich are attracted to areas with good housing and actually deterred by the existence of bad housing. The poor are attracted to bad housing and also positively to areas where good housing also exists. That is, it is assumed that all the parameters in (5.110) and (5.111) are positive. Then, the differential equations are

$$\dot{H}_i^1 = \rho^1 (\frac{h_1}{\lambda_1} \Sigma_j T_{ij}^{11} - q^1 H_i^1) \qquad (5.112)$$

and

$$\dot{H}_i^2 = \rho^2 (\frac{h_2}{\lambda_2} \Sigma_j T_{ij}^{22} - q^2 H_i^2) \qquad (5.113)$$

using obvious extensions of the parameter definitions introduced earlier.

The equilibrium conditions can then be written

$$\frac{h_1}{\lambda_1} \sum_j \frac{B_j^1 E_j^1 (H_i^1)^{\alpha_1^1} (1 + H_i^2)^{-\alpha_2^1} e^{-\mu c_{ij}}}{\sum_k (H_k^1)^{\alpha_1^1} (H_k^2)^{\alpha_2^1} e^{-\mu c_{ik}}} = q^1 H_i^1 \qquad (5.114)$$

and

$$\frac{h_2}{\lambda_2} \sum_j \frac{B_j^2 E_j^2 (H_i^2)^{\alpha_1^2} (H_i^2)^{\alpha_2^2} e^{-\mu c_{ij}}}{\sum_k (H_k^1)^{\alpha_1^2} (H_k^2)^{\alpha_2^2} e^{-\mu c_{ik}}} = q^2 H_i^2 \qquad (5.115)$$

The analysis can now proceed along the lines of previous ones, with α_1^1 and α_2^2 respectively playing the role of 'α' factors in the two sets of equations. The new feature introduced here, however, is the strong coupling between the equations through the appearance of one of the main variables of one equation as a factor in the attractiveness term of another, and this can be expected to generate interesting bifurcation behaviour, especially because of the negative coefficient on H_i^2 in Equation (5.114). What is necessary is to investigate what the joint equilibrium of the two systems look like in particular cases and whether there are sudden changes in pattern at critical parameter values.

147

All of these residential models illustrate another feature of complexity which we have as yet neglected throughout: we have so far assumed in both retail and residential models that 'cost' terms like k, q or r are constant. Much more theory can be added to this from urban economics to build in any known spatial variation. This does not change the argument in principle. A term like q above becomes q_i and may in part be explicitly a function of terms involving model variables like density (which would include H_i). If this was the case, then new non-linearities would be introduced and hence new bifurcation behaviour.

5.4.5 The Lowry model

The basic equations for an aggregate Lowry model were spelled out in Section 5.2.4. We noted the strong coupling of the residential and retail models through the fact that the field around population generated service employment, and the field around total employment generated the population. In the Lowry model, there is an implicit simple assumption about housing: that it develops as needed by the population location model. We could if necessary incorporate this assumption by taking $H_i = P_i = W_i^{res}$. The main point to be noted briefly here is that the analyses of possible bifurcation behaviour applied separately above to retail and residential models can be applied to them as Lowry model components. The effect of the coupling would be that any jump, for example, in one would be transmitted to the other. This secondary impulse could well in some cases take the variable concerned through a critical point and cause further jumps. This is rather reminiscent of the 'domino effect' discussed by Isard and Liossatos (1978). The presence of these secondary and reverburating phenomena will make jumps all the more difficult to interpret in linked models such as the Lowry model.

5.4.6 Disaggregated models

We have already anticipated the discussion of
disaggregation somewhat in the section on residential location
above. However, we can usefully return to the principles outlined in Section 5.2.5 as applied to the retail model. We will
consider the model given by Equations (5.13) and (5.14) with
attractive functions given by Equations (5.16). Take the case
where there are only two types (or orders) of good, for
simplicity. The basic equations are repeated below for
convenience.

$$S_{ij}^g = A_i^g e_i^g P_i \hat{W}_j^g e^{-\beta^g c_{ij}} \qquad (5.116)$$

$$A_i^g = 1/\sum_k \hat{W}_k^g e^{-\beta^g c_{ik}} \qquad (5.117)$$

we have

$$W_j^* = W_j^1 + W_j^2 \qquad (5.118)$$

so that

$$\hat{W}_j^1 = W_j^{*\alpha_1^1} W_j^{1\alpha_2^1} \qquad (5.119)$$

and

$$\hat{W}_j^2 = W_j^{*\alpha_1^2} W_j^{2\alpha_2^2} \qquad (5.120)$$

The following balancing conditions can be assumed to apply in
each sector, allowing the possibility of a different 'k' in
each:

$$\begin{array}{cc} D_j^1 = k_1 W_j^1 &) \\ &) \\ &) \qquad (5.121) \\ D_j^2 = k_2 W_j^2 &) \end{array}$$

The dynamical analysis can then be carried out for each sector
in turn - setting up $D_j^{1(1)}$, $D_j^{1(2)}$, $D_j^{2(1)}$ and $D_j^{2(2)}$ curves and
lines and calculating the derivatives to get the shapes of the
revenue curves. These calculations are relegated to Appendix 4.
It is shown that the results are similar to those of the
aggregate case but with α_2^1 and α_2^2 playing the role of the old
single α parameter for each sector in turn. If either is

Chapter 5

greater than 1, then jumps are possible in that sector; otherwise not. This leads to some interesting speculations along the following lines. Define

$$\alpha_1^1 + \alpha_2^1 = \alpha^{(1)} \tag{5.122}$$

$$\alpha_1^2 + \alpha_2^2 = \alpha^{(2)} \tag{5.123}$$

so that $\alpha^{(1)}$ and $\alpha^{(2)}$ are playing the role of α parameters in an overall sense. Then, if sector 1 is taken as representing consumable goods and sector 2 durable goods, we might expect

$$\alpha^{(2)} >> \alpha^{(1)} \tag{5.124}$$

and for $\alpha^{(1)}$ to be of the order of 1. This suggests, examining (5.122) and (5.123), that α_2^1 is likely to be less than 1 while α_2^2 has the possibility of being greater than 1. Generally, therefore, there is much more likelihood of jump behaviour for higher order goods, and this is in accordance with our intuition. The more complicated the system becomes, the more orders of goods there are, and then the same argument shows that it is less likely that there will be jumps in the lower orders.

5.4.7 Ecological analysis

In Chapter 2, we made much use of differential equations from ecology, either prey-predator Lotka-Volterra equations or competition-for-resources (CR) equations to illustrate bifurcation phenomena. Here, we can make direct use of the analogy to get new insights into locational models. It turns out that the kinds of equations we have been building for W_j behaviour in the retailing system, for example, have a topological similarity to the CR equations. This is in a sense not surprising intuitively, once the analogy has been spotted: our suppliers of shopping centres are competing for consumers' cash, which is in fixed supply. The complexities of our problem arise from the large number of dimensions involved (defined by number of zones, and equivalent to the number of species being modelled).

In Chapter 2, we presented an account, following Hirsch and Smale (1974), of the competition-for-resources model for two species. We now generalise the equations to N species and apply them to the retail model case. It will remain true that we can only examine the results graphically for the two-dimensional case, but we offer some speculations on how they might generalise.

We begin by repeating Equation (5.38) for convenience here:

$$\dot{W}_j = M_j(W_1, W_2, \ldots, W_N)W_j \tag{5.125}$$

For the retail model

$$M_j(W_1, W_2, \ldots, W_N) = \varepsilon \left[\sum_i \frac{e_i P_i W_j^{\alpha-1} e^{-\beta c_{ij}}}{\sum_k W_k^\alpha e^{-\beta c_{ik}}} - k \right] \tag{5.126}$$

The conditions which Hirsch and Smale specify as being characteristics of this problem as as follows:

(1) If any W_k increases, there will be a decrease in W_j, $j \neq k$:

$$\frac{\partial M_j}{\partial W_k} < 0, \quad k \neq j \tag{5.127}$$

This can easily be seen to be true for the retailing case, since

$$\frac{\partial M_j}{\partial W_j} = -\varepsilon \left[\sum_i \frac{\alpha e_i P_i W_j^{\alpha-1} W_k^{\alpha-1} e^{-\beta c_{ij}} e^{-\beta c_{ik}}}{(\sum_k W_k^\alpha e^{-\beta c_{ik}})^2} \right] \tag{5.128}$$

and this is obviously negative for $\alpha > 0$.

(2) W_j cannot expand indefinitely. This can be expressed by saying that there exists a constant, K say, such that

$$M_j < 0 \text{ if } W_j > K \tag{5.129}$$

We can see from Equation (5.126) that this holds for the retailing case. The summed term decreases as W_j increases, since the numerator contains a $W_j^{\alpha-1}$ term and the denominator a W_j^α term.

Chapter 5

This means that there must exist a W_j at which this summed term becomes less than k, which is the required result.

(3) If there is only one species (or centre), then it grows to a certain point but not beyond it. This can be expressed by saying that there exists an a_j such that

$$M_j(0,0, \ldots, W_j, 0, \ldots, 0) \begin{cases} >0 \text{ if } W_j < a_j \\ <0 \text{ if } W_j > a_j \end{cases} \qquad (5.130)$$

This can easily be proved for the retail case:

$$M_j(0,0, \ldots, W_j, 0, \ldots, 0) = \varepsilon \left[\frac{\sum_i e_i P_i}{W_j} - k \right] \qquad (5.131)$$

Thus:

$$M_j > 0 \text{ if } W_j < \frac{\sum_i e_i P_i}{K} \qquad (5.132)$$

and vice versa.

The results for the two zone (or two species) case were presented in Chapter 2, Section 2.3.4 and particular in Figure 2.24. This shows the case where there are two possible stable equilibrium points, one with both W_j's (reinterpreting the names of the variables) non-zero and one with one of them zero - the coexistence and one species-annhilation solutions. We saw that they are divided by a separatrix which passes through the saddle point Q. We can now see that, if the system is disturbed from equilibrium in such a way that it crosses the separatrix, then it will return to a new and different equilibrium point. This is an example, new for the urban context, of a different kind of bifurcation.

Broadly speaking, we can expect the two-dimensional Hirsch and Smale results to generalise to N dimensions. The first point to note relates to uniqueness: if an equilibrium point exists with all the W_j's non-zero, then it is thought to be unique.[3] There will be many other stable points with one or more W_j's zero and we can expect these to be divided by separatrices which generate the kind of bifurcation behaviour we have been outlining in two dimensions. But it is obviously difficult to be more specific. We can also note that our previous analyses, which established conditions for zones to be in the DP state, give some indication of when the all-non-zero point may not exist: when there is at least one zone in the NDP state, and so we learn something about the associated existence theorem.

More generally, these new insights do suggest that there is something to be learned from seeking analogies with certain ecological models and we will pursue this further in Chapter 8 below.

Notes

(1) It could be argued, as for the retail model, that the addition of a balancing factor like (5.27) makes it quasi-doubly-constrained.

(2) Here, and in the following argument, W_j refers to the *equilibrium* value. That is, we are restricted to W_j given by (5.73).

(3) There is one worry which needs to be resolved here: if the argument about the possibility of multiple inflexions in the revenue curve, presented above, is borne out in practice, then this seems to suggest the possibility of a situation in which there could be more than one point with all W_j's non-zero.

CHAPTER 6

BIFURCATION AT THE MESO-SCALE II: THE DYNAMICS OF URBAN SPATIAL STRUCTURE

6.1 Introduction

In the preceding chapter, we set up dynamical equations for some important urban subsystems and carried through in depth an analysis of the stability of equilibrium points. This forms the basis of a 'comparative static' approach to the analysis of change, but with more than the usual interest because of the possibilities of bifurcation. In this chapter, we extend the argument to systems which are not in equilibrium. In Section 6.2 we discuss bifurcation and disequilibrium and the effect of fluctuations on systems. The argument is extended in Section 6.3 to control theory. In Section 6.4, a number of results are drawn together and it is shown that an alternative formulation of central place theory can be developed. Some directions for future research are considered briefly in Section 6.5.

6.2 Disequilibrium, fluctuations and bifurcation

6.2.1 Introduction

In Section 5.4, we concentrated on the analysis of equilibrium states: their multiplicity, stability and the way in which these characteristics change with parameter values. In one sense, the extension of the argument to deal with systems in disequilibrium is simply an acknowledgement of the need to integrate the various differential equations numerically, starting from initial conditions which represent a system state not in equilibrium, and to explore the trajectories which result. We saw from Chapter 2 that such trajectories will be strongly influenced by the position of the equilibrium points, in different ways according to whether they are stable, unstable or saddle points. The trajectories can only be charted geometrically when the phase space is two dimensional

and typically, of course, this is not the case for the
interesting examples of urban spatial structure at the meso
scale. We therefore use this section to explore methods and
formulations which differ from those presented above but which
also illustrate the idea of disequilibrium. The first
example, the work of the Brussels school, builds on the ideas
of Prigogine and others in physics and chemistry (*cf*. Nicolis
and Prigogine, 1977). This places an emphasis not only on disequilibrium, but also on the idea of the system being subjected
to random fluctuations. In the second example we look more
generally at accounting equations and discuss how ideas on
bifurcation can be applied to them. The concept of fluctuations
can be applied in that context also. A number of concluding
comments are made in Section 6.2.4.

6.2.2 Order from fluctuations : the Brussels school

Prigogine and others have produced important results about
physical and chemical systems which, they argue, demonstrate
the production of 'order from fluctuations'. Their ideas will
be discussed in more detail in Chapter 8. A number of
research workers in Prigogine's department, however, are now
applying these ideas to urban modelling and these fit better
into the structure of this book at this point.

The main ideas can be sketched very briefly, for systems
in general, as follows. The base-level state is the
'thermodynamic equilibrium'. This is the state attained by a
closed system (into or out of which there is no flow of
energy or matter). For an open system, there are flows of
energy or matter which can take the system to a state 'far from
equilibrium'. The particular systems of interest in this context are described by equations which are non-linear and which
include representations of interactions or feedbacks between
different elements of the system. These generate possibilities
of bifurcations and some states of the system appear unexpec-
tedly, at critical parameter values, as ordered structures - the
'order' being in either time or space or both.

Note one important point at the outset: the notion of
'equilibrium' here is somewhat different to the one we have
used earlier. It is very much a base-level equilibrium for a
closed system. The equilibria we described in Section 5.4 are
more like steady-state equilibria for a dynamical system. We
have focussed on the changing nature of this equilibrium state
as parameters change, and this is equivalent to having a supply
of energy or matter driving the system to new states. Thus,
there is no very basic difference in approach.

The Brussels' workers, however, do add stochastic terms.
In one respect this is simply useful as additional realism,
since such effects are obviously present in urban systems.
There may be another more important consequence, however. We
noted at one point in the previous section that there were
multiple   solutions available for the whole system, distinguished in that example by the number of W_j's which took zero
values. The addition of random fluctuations to the numerical
integration of differential equations, of the kind we have been
studying, may serve to chart the range of alternative routes
for system development, selecting in different ways, at
different times and for different runs of the model, the particular equilibrium state which dominate particular trajectories.
There are obviously some important issues for further research
here.

Meanwhile, as a start, we outline the kinds of models
constructed in Brussels and compare them with the ones outlined
above. The following account is based on Allen *et al*. (1978)
and Allen and Sanglier (1979). They have developed two main
models, one inter-urban - representing systems of cities - the
other urban, representing the location of activities within a
city. We begin the the inter-urban model.

In the following account, we pick out the bare essentials
of the Brussels' model without offering the detail justifications
for functional forms which the authors provide, though some of
these issues will also be taken up later. In the inter-urban

model, the main form of differential equation used is based on
the Verhulst form of logistic growth model for populations
(*cf.* Chapter 2). For a population, x, this can be written

$$\frac{dx}{dt} = bx(P - x) - dx \qquad (6.1)$$

where b is a birth-rate, d is a death-rate and P is a measure
of the population-carrying capacity of the area. The term
P - x, in effect, modifies the birth-rate to provide an upper
bound to growth - to 'prevent' exponential growth.

The form in which Allen *et al.* apply this to a city -
labelled i and within a system of cities - is

$$\frac{dx_i}{dt} = bx_i \left[P + \Sigma_k R^k S_i^k - x_i \right] - mx_i \qquad (6.2)$$

where b now represents the rate of birth and immigration, m the
rate of death and emigration and P is a carrying capacity as
before. It is argued, however, that this capacity can be
expanded by the presence of new economic activities: S_i^k is the
amount of employment in such activities, k, at i, and R^k a set
of constants measuring their impact. It is then necessary to
add equations which describe the growth of these new economic
activities and this also takes a logistic form:

$$\frac{dS_i^k}{dt} = \alpha [E_i^k - \gamma^k S_i^k] S_i^k \qquad (6.3)$$

The 'capacity' term is E_i^k (in the notation of Allen and
Sanglier, 1979) which is the amount of employment potentially
generated at i by the demand for sector k. α and γ^k are
suitable constants. If other constants p^k are then defined, E_i^k
can be written

$$E_i^k = p^k D_i^k \qquad (6.4)$$

The problem, then, of course, is to find D_i^k, the demand for k
at i. At this point, the argument becomes rather complicated.

The demand at i generated by residents of j is initially taken to be

$$D_{ij}^k = \frac{\epsilon^k x_j}{(p_{ij}^k)^e} \tag{6.5}$$

so that

$$D_i^k = \sum_i D_{ij}^k = \sum_j \frac{\epsilon^k x_j}{(p_{ij}^k)^e} \tag{6.6}$$

Here, ϵ^k and e are constants. The former is the average demand from an individual for the product of k, though modified by the price as we will see. p_{ij}^k is the price of a unit of k, produced at i, to a resident of j. This is assumed to have a spatial variation as follows:

$$p_{ij}^k = p_i^k + \phi^k d_{ij} \tag{6.7}$$

where p_i^k is the production cost at i and $\phi^k d_{ij}$ represents the amount to be added, and charged to the customer, as a result of transportation costs; d_{ij} is the inter-zonal distance and ϕ^k a set of constants representing unit transport costs by sector. Thus, demand at i, from j, is supposed to decrease with distance in a Christaller-like way: Equation (6.5) together now with (6.7) represents the idea of a Christaller demand cone.

However, the argument cannot stop there about demand: it is necessary to add the effects of competition of other zones. Define A_{ij}^k as the attractiveness of i for residents of j, for sector k, relative to other zones. This is taken as being proportional to a function of the number of facilities (ie. k sectors) available at i - to allow for multipurpose trips and so on - and inversely proportional to the price of the good at j. Thus

$$A_{ij}^k = \frac{(1 + \rho n_i)}{p_{ij}^k} \tag{6.8}$$

Chapter 6

where n_i is the number of facilities at i, and ρ is a constant. The factor which is then added to D_{ij}^k to represent relative attractiveness is

$$\frac{(A_{ij}^k)^I}{\sum_{i'}(A_{i'j}^k)^I} \tag{6.9}$$

I is a constant which measures the sensitivity of the population to the differences represented in A_{ij}^k: if the population is very sensitive, I is high, and vice versa.

The final step in the argument is to modify the price of a unit of k at i, p_i^k, to allow for economies of scale:

$$p_i^k = \sigma^k + \frac{\Delta^k}{\gamma^k[aS_i^k - b(S_i^k)^2]} \tag{6.10}$$

σ^k, Δ^k, a and b are suitable constants. In this formulation, the optimum size of a plant at i is given by

$$S_i^k = a/2b \tag{6.11}$$

above which diseconomies of scale set in. Thus, D_{ij}^k can be put together using Equations (6.5), (6.7), (6.8), (6.9) and (6.10) as

$$D_{ij}^k = \frac{x_j \epsilon^k \left[\sigma^k + \dfrac{\Delta^k}{\gamma^k[aS_i^k - b(S_i^k)^2]} + \phi^k d_{ij}\right]^{-e-I}(1 + \rho n_i)^I}{\sum_{i'}(1 + \rho n_i)^I \left[\sigma^k + \dfrac{\Delta^k}{\gamma^k[aS_{i'}^k - b(S_{i'}^k)^2]} + \phi^k d_{i'j}\right]^{-I}} \tag{6.12}$$

and D_i^k can be obtained from (6.6).

A quick glance at Equation (6.3) shows that the equilibrium conditions for S_i^k could be written as

$$p^k \sum_j \frac{\varepsilon^k x_j \left[\sigma^k + \dfrac{\Delta^k}{\gamma^k \{aS_i^k - b(S_i^k)^2\}} + \phi^k d_{ij}\right]^{-e-I} (1 + \rho n_i)^I}{\sum_{i'} (1 + \rho n_{i'})^I \left[\sigma^k + \dfrac{\Delta^k}{\gamma^k \{aS_{i'}^k - b(S_{i'}^k)^2\}} + \phi^k d_{i'j}\right]^{-I}} = \gamma^k S_i^k \tag{6.13}$$

Equation (6.2) shows that the equilibrium condition for the population variable x_i can be written

$$P + \sum_k R^k S_i^k = x_i \tag{6.14}$$

Equation (6.13) has been written out in full to expose all the complexities involved in seeking an analytical solution for S_i^k, or indeed any geometrical insights. The main non-linearities arise in three ways, all difficult to handle:
(1) the way in which the S_i^k's occur in the economies-of-scale term, and the way this term appears in the numerator, thus including every S_i^k, term, $i' \neq i$, in each i-k equation;
(2) the number of facilities, n_i, depends on the pattern of non-zero S_i^k's, varying over k for each i; (3) each i-k equation involves all the x_j's and this provides a strong coupling to each x_j equation, each of which involves all the S_i^k's.

Equation (6.14) shows that the equilibrium value for each x_i depends on the equilibrium values of the S_i^k's, and so these values can only be investigated as the solution to strongly coupled sets of equations.

It is no wonder in the face of this that Allen *et al.* concentrate on exploring the structures which emerge in simulation runs rather than trying to analyse the nature of the equilibrium points which shape such runs. The way the simulation proceeds is as follows. Initial values are assigned to the populations which are equilibrium values in the case where all the S_i^k's are zero. The spatial system is a set of zone centroids which form a triangular lattice. It is assumed that increasing values of k represent increasing orders in an

Chapter 6

economic hierarchy, and this is expressed by increasing values of γ^k with k. γ^k plays the role of a threshold - as does the parameter k in the Section 5.4 (of Chapter 5) model of course.

The equations are then integrated numerically over a succession of discrete time intervals. In the early stages, small amounts of S_1^1 are added to random points (and there are sufficient time intervals to ensure that each point is given an opportunity to develop as a 1-centre). This leads to the growth of sector 1 at a number of points. When these have reached some kind of equilibrium, sector 2 is introduced on the same basis, and so on. In the experiments reported, three sectors were used.

This procedure generates hierarchically-structured spatial patterns: the actual pattern varies between runs, but there are some regularities in the relative numbers of centres in each rank. The fact that the spatial pattern varies supports the hypothesis mentioned earlier that there are multiple stable equilibrium states to choose from, which shape a trajectory even when the system never achieves steady state, and that the particular path taken by the system as a whole is then determined by the fluctuations. The role of the fluctuations is important in the sense that, given an initial state, the system without fluctuations will tend to stay nearest to the equilibrium point that it starts from. The number of alternatives which are achievable will depend on the magnitude of the fluctuations. What is not clear is the point (in time) at which some parameters have critical values nor the form of the underlying bifurcations. Further, our previous analysis of the simpler case of Section 5.4 suggests that critical parameter values are likely to exist, but also that separatrix crossing jumps might be possible because of the fluctuations. The interpretation of the results of the Brussels' model, therefore, remains an interesting research question.

Next, we consider the urban model. The spatial basis of this model is a square lattice and the basic differential

162

equation which has been used is similar to that used in Section 5.4, rather than the logistic equation. For a variable x this takes the form

$$\frac{dx}{dt} = \Theta(D - x) \tag{6.15}$$

where Θ is a parameter and D is a capacity. As we noted in Chapter 2, this changes the initial growth pattern but not, essentially, the form of the underlying equilibrium points at any particular time. The form of the overall model is not entirely clear in the presentation in Allen *et al.* (1978) and so here we concentrate only on what appear to be the main points rather than giving a full connected statement of the model.

Initially, three economic sectors are defined: export (X_i), services (S_i) and industrial services to export industries (V_i). Later a fourth 'mixed' sector, Z_i, is defined. A population sector (P_i) is also defined. There are basically two forms of differential equation. One applies only to the export sector, the other to services, services to export and to population. The export equation takes the form, for each zone i,

$$\frac{dX_i}{dt} = \Theta^E(D_i^E - X_i) \tag{6.16}$$

which can be written

$$\frac{dX_i}{dt} = \Theta^E\left(\frac{D^E F_i}{\sum_k F_k} - X_i\right) \tag{6.17}$$

where F_i is a measure of the attractiveness of zone i for the location of export industries within the city, Θ^E is a constant and D^E is total demand for this product. The other sector equations are a modified form of (6.16); their equivalent of the D_i^E term is built up from demands from all the zones of the city:

$$D_i^S = \sum_j D_{ij}^S \tag{6.18}$$

Chapter 6

$$D_{ij}^S = b_{ij}P_j \tag{6.19}$$

Here, D_i^S is the total demand for services at i and this is built up from demands from the population in all zones j through a set of coefficients, b_{ij}. These can be taken as a constant times relative attractiveness term F_{ij}^S. (The notation of Allen *et al.* has been modified slightly in this account because they do not distinguish explicitly constants and demand terms which are obviously different.) Then, the service equation can be written

$$\frac{dS_i}{dt} = \theta^S(\sum_j \frac{F_{ij}^S \epsilon^S P_i}{\sum_k F_{ij}^S} - S_i) \tag{6.20}$$

A similar argument for the V_i sector produces

$$D_i^V = \sum_j D_{ij}^V \tag{6.21}$$

$$D_{ij}^V = a_{ij}X_j \tag{6.22}$$

and

$$\frac{dV_i}{dt} = \theta^V(\sum_j \frac{F_{ij}^V \epsilon^V X_j}{\sum_k F_{ik}^V} - V_i) \tag{6.23}$$

using an obvious extension of the earlier notation. The mixed sector, Z_i, arises from a mixed form of the two kinds of equations:

$$\frac{dZ_i}{dt} = \theta^Z(a^Z \sum_j \frac{F_{ij}^S P_j}{\sum_k F_{ik}^S} + b^Z \frac{F_i}{\sum_k F_k} X_i - Z_i) \tag{6.24}$$

where a^Z and b^Z are constants which determine the relative importance of population and export industries in the locational behaviour of this sector together with its scale. The residential model is built up as

$$\frac{dP_i}{dt} = \Theta^R (\sum_j \frac{F_{ij}^R D_j}{\sum_k F_{ik}^R} - P_i) \tag{6.25}$$

and in this case it is sometimes useful to write down an equation for its components P_{ij}:

$$\frac{dP_{ij}}{dt} = \Theta^R (\frac{F_{ij}^R}{\sum_k F_{kj}^R} \cdot D_j - P_{ij}) \tag{6.26}$$

In these equations, D_j is total employment at j which could be written as

$$D_j = X_j + S_j + V_j + Z_j \tag{6.27}$$

provided all the terms on the right hand side are given in employment units, or with appropriate coefficients added otherwise.

Obviously the detailed form of the model turns on the 'location functions' F_i, F_{ij}^S, F_{ij}^V and F_{ij}^R. In the rest of the paper, Allen *et al.* discuss the form of F_{ij}^R in detail but not the others. Indeed, in the model whose results are reported, the focus is entirely on residential patterns and the economic sector is treated in aggregate with a single variable X_i being used to represent jobs at i. It would in fact be more consistent, in the light of Equation (6.27) to use D_j for this, and this is what is done below. It is also not clear from the text how many zones are allowed to have jobs: that is, the extent to which they are concentrated in the centre of the study area. Here, the formulation as presented is modified to allow for all zones having jobs. The attractiveness function can then be taken as

$$F_{ij}^{Rk} = \frac{G_i^k \left[\frac{u_i^k}{C + u_i^k} \right]}{\left[B1^k + AZ^k (P_i^k + P_i^{k'}) + CL^k d_{ij}^2 \right]} \tag{6.28}$$

which obviously needs some explanation.

Chapter 6

k is now, used as an index to distinguish person types and the purpose of the prototype model is to show the effects of interaction and competition between two such types. In a rather ingenious notation, the two population types are labelled as k and k' with the convention that when both appear in an equation, if k = 1, then k' = 2, and vice versa. G_i^k measures the effect of neighbourhood zones and is given by

$$G_i^k = \sum_{h=h_1}^{h_2} (A12P_h^k + A13P_h^{k'}) \tag{6.29}$$

where the constants A12 and A13 measure the relative importance of the different groups to k groups and the summation is over neighbouring zones. U_i^k is a utility term given by

$$U_i^k = \exp\left[AV1(P_i^k - P_i^{k'})(P_i^k + P_i^{k'})\right] \tag{6.30}$$

AV1 is a constant. The argument for this is that, for k, it will be highest when there is a large k population in that zone (and higher still if there is a large total population) and so this can be the basis of segregation effects. The second two terms in the denominator measure the disbenefits of crowding and the costs of the journey to work to j respectively. In some versions of the model, a term ED_j is also added (for a suitable constant E) to allow for the competition for land between residential and another economic activity. The U_i^k term appears in the form $\dfrac{U_i^k}{C + U_i^k}$ in Equation (6.28) to ensure that its effect has an upper bound.

This model is obviously disaggregated relative to (6.25) and can be written out in full as

$$\frac{dP_i^k}{dt} = \Theta^R\left[\sum_j \frac{F_{ij}^{Rk}D_j}{\sum_{i'} F_{i'j}^{Rk}} - P_i^k\right] \tag{6.31}$$

It is easy to see that the equilibrium condition is

166

$$\sum_j \frac{F_{ij}^{Rk} D_j}{\sum_{i'} F_{i'j}^{Rk}} = P_i^k \qquad (6.32)$$

but that given F_{ij}^{Rk} in (6.28) and the definitions of all the component pieces of that attractiveness factor, then the non-linearities in the P_i^k variables are again impossible to handle analytically. Both k and k' are clearly involved in each equation, but so are the P_i^k's for each zone because of the $\sum_{i'} F_{i'j}^{Rk}$ term in (6.32). So once again, an analytical search for solutions is likely to be unfruitful. This case is, however, simpler than the inter-urban example, and it may be possible to obtain numerically plots of

$$\sum_j \frac{F_{ij}^{Rk} D_j}{\sum_{i'} F_{i'j}^{Rk}} \qquad (6.33)$$

and then to use a method analogous to that of Section 5.4.

In the simulation runs, a different procedure is used to introduce fluctuations. It is not entirely clear whether the distribution of jobs is given initially and fixed throughout the simulation, or whether there is some growth process from various initial conditions. In either case, it is not clear what the initial conditions are. However, we will assume that a satisfactory procedure is adopted. The random effects are then introduced as follows:

$$D_j^N = D_j^A(1 - \lambda) + \lambda D_j^A G_j \qquad (6.34)$$

where D_j^A is the spatial distribution of jobs at a particular time interval and D_j^N is a new distribution for the next time interval in the numerical integration of the equations. λ is a parameter and G_j is a random number between 0 and 1 scaled so that

$$\sum_j G_j = 1 \qquad (6.35)$$

Chapter 6

The results of the simulation runs do show different patterns of residential segregation developing for a wide series of different parameter values. However, because the data is essentially hypothetical, the main utility of the present model is theoretical. In the next subsection, we briefly review the possibility of working with more general sets of equations and reserve our final comments on the Brussels' model to a context which can include this other approach and this is done in the final Subsection 6.2.4, below.

6.2.3 The use of kinetic equations

We will see in Chapter 8 that one of the most common examples of bifurcation behaviour in non-linear systems arises in chemistry in equations describing kinetics of reactions. These kinds of equations also occur in urban modelling and it is useful at this stage to add a note on the differences between such an approach and the ones describe above and to identify an important research topic for future research on bifurcation.

In the approaches in Sections 5.4 and here, the focus is essentially on growth to an equilibrium position which is embedded in the equations. For example

$$\frac{dx_i}{dt} = \epsilon\left[D(x_i) - x_i\right]x_i \tag{6.36}$$

In cases where the main source of change arises from the transitions of elements from one state to another, it will be more appropriate to write the differential equations in terms of transition coefficients:

$$\frac{dx_i}{dt} = \sum_j (a_{ji}X_j - a_{ij}X_i) \tag{6.37}$$

where a_{ij} is the transition rate per unit time from state i to state j. In suitable cases, where a Markovian assumption can be made, and with a suitable definition of a_{ii}, such equations can be written

$$\frac{dx_i}{dt} = \sum_j a_{ji} x_j \qquad (6.38)$$

The modeller now has a different focus. In relation to
Equation (6.36) the emphasis is on representing the demand term
$D(x_i)$ and modelling the progress towards the equilibrium state

$$D(x_i) = x_i \qquad (6.39)$$

(which, of course, may never be reached) in a situation where
many exogenous variables and parameters may be changing over
time. With Equation (6.37) the modeller's task is to represent
the transition rates, a_{ij}. The interest in the present context
is that there is every reason to think in a number of cases
that these will contain components, for example representing
push and pull factors, which are rather like the attractiveness
terms which play a prominent role in the other kinds of models.
In particular, they are likely to contain x_i (and x_j, $j \neq i$) in
non-linear form which means that the equilibrium condition can
be written as

$$\sum_j a_{ji}(x_1, x_2, \ldots, x_n) x_j = x_i \sum_j a_{ij}(x_1, x_2, \ldots, x_n) \qquad (6.40)$$

and we can expect these points to have interesting bifurcation
properties.

Equations of the form (6.37) and (6.38) have been developed
in the context of residential mobility (see for example,
Ginsberg, 1973, 1978 and, for a review, Wilson, 1979-C),
migration (Gleave and Cordey-Hayes, 1977, Varaprasad, 1979) and
in transport (Tomlin, 1969). In most of these cases, the main
state variables are doubly subscripted and the equations take
the form

$$\frac{\partial T_{ij}}{\partial t} = \sum_{pq} a_{pq,ij} T_{pq} \qquad (6.41)$$

Further, there is no reason why stochastic terms should not be
added to such a model. It could also be used to model
disequilibrium by numerical integration from suitable initial
conditions.

Chapter 6

6.2.4 Concluding comments

In this section, we have concentrated mainly on the
Brussels' models as a demonstration of how stochastic variables
can be incorporated into dynamic models. These models can be
criticised in details. For example, the authors of the main
papers (Allen *et al.*, 1978) discuss many features of urban and
regional systems which they would like to see built into models
but then do not put these together consistently. For example,
they sometimes define economic sectors according to order in a
hierarchy and at other times in relation to export/service/
mixed distinctions. These two kinds of concepts obviously need
to be forced together and should have something like an underlying input-output model as an aggregate basis. It would also
be difficult to find many of the 'constants' which are defined
for the models in practice.

Like all modellers, the Brussels' school face a complicated
choice between making the models as realistic as possible and
making them sufficiently simple for analysis and computation to
be possible. There is an argument for saying that they have
fallen between two stools: the models are not realistic in
that some of the mechanisms are too simple and turn on constants
which are impossible to estimate and which are made to do too
much work; on the other hand, the attractiveness functions
used make it very difficult to gain any analytical insights into
the nature of the models. It would be an interesting experiment,
therefore, either to build simpler functions into their models,
or to build stochastic variation into the models described in
Section 5.4.

What we should emphasise, however, is the important nature
of the additions offered by the Brussels' work. Allen *et al.*
(1978) discuss three types of fluctuations: (1) in state
variables; (2) in structural innovation; and (iii) in
environmental variables such as parameters. The urban model
offers examples of the first kind, the inter-urban model,
examples of the second (though in a very simple form and we

170

return to this issue in Section 6.5 below). They have not yet experimented with the third kind. They also show how to obtain results for systems which are never in equilibrium but whose behaviour is governed by underlying (multiple) stable equilibrium points. In effect, they provide a method, through the use of fluctuations, for sampling the trajectories taken by the system as a whole through the multiple stable states available to it.

It is certainly true that these innovations could be added to other types of dynamic models. These could be of the style of Section 5.4 or the kinetic equations discussed briefly in Section 6.2.3. Indeed, the methods of both 5.4 and this section could be applied to any kinds of dynamical system model. Another possible line of research, where the examination of bifurcation properties could be interesting, would be the so-called 'systems dynamics' models of Forrester (1968) and others.

6.3 Control theoretic formulations
6.3.1 Introduction

The next possible extension of these methods is a very difficult one. When dynamical systems models are used in a planning context, it will be the case that some variables are controllable. Design values of such variables are sometimes obtained in a static context using mathematical programming, but it is more difficult to optimise the time pattern of such variables. This is partly because such problems are difficult to formulate, but also because, when this can be achieved, they are difficult to solve in practice. In this section, we introduce briefly an example of such a problem, again in the retail sector, and then in Section 6.3.3 briefly explore a wider possible range of applications. The potentially interesting feature of these problems in the context of this book is that parts of the system can be represented by differential equations and these function as constraintequations in the overall control problem if the solutions to these equations are in the neighbourhood of critical points.

Chapter 6

6.3.2 A control problem in shopping centre location

The obvious thing to try to control in the shopping centre problem is the size and location of the centres. In order to illustrate the nature of the control problem, we will assume that some of the centres can be controlled by a planning authority and some not - the latter being governed by the usual supply-side differential equations. Suppose the numbering can be ordered in such a way that centres 1, 2, ..., R are governed by the usual 'entrepreneurial' differential equations and R + 1, ..., N are to be controlled. All other variables are as usual for the aggregate problem. It is shown in Appendix 1 that a suitable objective function is to maximise consumers surplus. Let this be $J(W_1, W_2, \ldots, W_N)$ given by:

$$J(W_1, W_2, \ldots, W_N) = -\sum_{ij} S_{ij}(\log S_{ij} - 1 + \alpha \log W_j - \beta c_{ij})$$

(6.42)

Then the control theory problem can be formulated as

$$\max_{\{W_{R+1}, \ldots, W_N\}} z(W_{R+1}, \ldots, W_N) = \int_0^T J(W_1, W_2, \ldots, W_N) dt$$

(6.43)

for some interval (0,T), subject to the constraints:

$$\dot{W}_j = \varepsilon\left[\sum_i \frac{e_i P_i W_j^\alpha e^{-\beta c_{ij}}}{\sum_k W_k^\alpha e^{-\beta c_{ik}}} - kW_j\right]$$

(6.44)

This is a problem which is solvable in control theory using a procedure known as the ε-method. This has been applied to the above problem by Phiri (1979) for a system of eight zones in which two of the shopping centres are controlled. There may be some difficulties in applying it to larger numbers of zones because of the amount of computation time involved.

6.3.3 Other possible applications

The most obvious extensions are to other cases where, for
a class of variables, it makes sense to control some of them
and to let others be governed by well-defined differential
equations. The most obvious example of these discussed so far
is residential location. The only difficulty in principle in
achieving this lies in the number of variables to be handled.
It would also be possible to link this with partial control of
some service centre variables.

A possible problem with the application of these methods
lies in the bifurcation properties of the differential equations
which are acting as constraints. It is customary in a control
problem over time to make exogenous assumptions about the
behaviour of some of the parameters or variables - like e_i, P_i
or c_{ij}. Changes in these variables could then cause jumps
which may be difficult for the control algorithms so far
available.

6.4 Integrated approaches : towards a new intra-urban central place theory

6.4.1 Introduction : the bases of central place theory

The most common application of central place theory is at
the inter-urban scale and we have seen one such modelling
application in the work of Allen *et al.* (1978) in Section 6.2.
Here, we concentrate more on the intra-urban scale, but the
problem is essentially the same: to develop a theory of the
development and spatial organisation of structures in a region
which have 'central' functions. At the inter-urban scale, this
leads to a concern with the hierarchical structure of systems
of cities; at the intra-urban scale, with the hierarchical
structure of facilities provided within the city, particularly
with respect to retailing.

There is a vast literature on central place theory. This
initial brief review is based mainly on Christaller's (1933)
original (in both senses) and masterly exposition. A more

173

recent review is that of Beavon (1977) and the earlier
literature in relation to intra-urban central place theory is
effectively covered by Berry (1967).

Christaller's scheme was developed under a set of highly
restrictive assumptions, though it should be noted that it is
clear from his own book (though not always from secondary
presentations) that he fully understood the nature of these
assumptions. Indeed, the approach of this section is similar
(and much of the book is in the same spirit: the models are
simpler than they will ultimately 'need' to be for empirical
effectiveness). However, the main argument of this section is
related to the nature of assumptions. It seems to be the case
that Christaller's more restrictive assumptions are forced on
him by the mathematical representation which he uses; the
kinds of models described earlier in this chapter use a
different one and it becomes possible to reformulate the
question of central place theory in such a way that the
assumptions made need not be so limiting.

In effect, this arises out of the way in which space is
treated: in traditional central place theory, space is handled
continuously (though boundaries appear which subdivide market
areas) while in the models described above, space is treated as
a set of discrete zones. Since they focus on *spatial interaction* between zones and the location of *activities*, we will
call the discrete zones models 'SIA models' for convenience.
This makes it easy not to have to assume that the population in
the area is uniformly distributed or that all jobs are at the
centre of a city or whatever. These kinds of assumptions are
forced on Christaller and others, as otherwise the mathematics
becomes intractable in the continuous space representations[1].
But this is to anticipate the argument: first, we review the
kinds of ideas which form the basis of traditional central
place theory.

In the broadest sense, the concepts of central place theory
can be applied in an obvious way at either inter- or intra-urban

scales and so in this brief account, we do not distinguish
unless it is absolutely essential. A spatially distributed
population is assumed to exist which demands goods and services.
'Centres' are places where organisations produce and/or supply
these. Most of central place theory is concerned with supply:
people travel to obtain the goods and services they require.
There is less detail in the theory on flows between production
or service organisations (which would involve a spatial representation of the kinds of transactions recorded in an input-
output model).

The theoretical basis therefore involves the theory of the
firm on the one hand and the theory of consumers' behaviour on
the other. The first of these will have to be sufficiently
broad to encompass the public sector in spatial economies
where this is appropriate. From the point of view of the firm,
it is usually assumed to be profit maximising (or perhaps cost
minimising for a public organisation); but more importantly,
taking account of the competition, for it to exist at all at a
location there has to be sufficient demand: this has to exceed
a certain threshold. For a given distribution of population,
this can be discussed in terms of the 'range of the good' and
whether within this range sufficient demand is generated to
sustain a firm of minimum size. The notions of 'market area'
and 'economies of scale' also have an obvious importance in
this context.

The theory of consumers' behaviour is usually based on
some notion of utility maximisation. People trade off the
utility of acquiring the good against the travel cost involved
in getting to the centre. 'Larger' (or 'higher order') centres
offer scale economies for the consumer because they may be able
to visit several kinds of establishments on one trip. Thus the
consumers are concerned with the utility acquired from the
goods, the cost of getting to centres, and with what we might
describe by our earlier concept of the overall 'attractiveness',
of centres. These trade-offs are much discussed in the literature,
for example in the recent review by von Böventer (1976).

175

Chapter 6

It will already be clear from the discussion so far that
there are two spatial scales involved: the underpinning theory
is based on micro-scale economics, whereas the main geographical results are at the meso scale: central places are
aggregates at a location of organisations supplying goods and
services. What is more, the meso-scale structure clearly
affects micro-scale behaviour because of both producers' and
consumers' scale economies, and because of externalities.

There is a basic measurement problem associated with
central place theory: how to combine individual activities at
a place into some measure of 'total size' - or perhaps again,
'attractiveness'. Many alternative solutions to this problem
are proposed in the literature and different ones are suitable
for different purposes. We have seen already with SIA models
that total floorspace for each good or service is sometimes
convenient, while Allen *et al.* (1978) have used, along with
many geographers, a count of the number of 'functions' (not
organisations or establishments) at a place. One way or
another, the aggregation problem relating the two scales has to
be resolved. It is probably easier in general to achieve this
for consumer behaviour than for firms.

The main results of central place theory are concerned with
hierarchical structure. There is an obvious sense in which a
hierarchy of places does exist: larger centres have a wider
range of functions than smaller ones, and a wider market area
for many goods. There is much controversy as to whether the
places at different levels of a hierarchy have typical sizes.
Some authors, such as Berry and Garrison (1958) have identified
discrete size distributions for example, while others such as
Beavon (1977) argue that the size distributions are continuous.
The hierarchy is usually simpler if counts of functions are
used as the measure: at higher levels, centres have more functions. Even then, however, it is not precisely the case that
all higher order centres have all lower order functions.

Centres are usually classified by 'type' in some way, perhaps based on the range of goods or services, but usually directly related to order in a hierarchy. The next theoretical question then concerns the number of centres of each type and where they are located. The latter question implies another about the (average?) spacing of centres. Authors such as Christaller (1933) and Lösch (1943) base their arguments on the micro theories mentioned earlier. At the meso scale, this generates centres of different orders with continuous market areas forming distinctive hexagonal patterns. These arise out of two basic notions: that the existence of a firm is determined by the range of its products and that consumers travel to the nearest centres at which goods or services are available. The Löschian pattern involves a relaxation of some of the assumptions and a more complicated structure, but one which essentially involves consumers going to the nearest centre also. Most applications described in standard texts are to systems of settlements, but the concepts can also be applied at the intra-urban scale.

The other assumption which generates the regularity of these patterns is that of the uniform distribution of population. It is clear, therefore, that when some of these assumptions break down - the population itself has a 'centre/ hierarchical' pattern - and there is 'dispersion' for various reasons in travel patterns, then the overall structure of centres is likely to be less neat. Nonetheless, it is argued that some of the relationships about numbers and spacing of centres at different levels in a hierarchy may still be maintained. In the model to be presented below, it is argued that because the assumptions can be relaxed (because of the different mathematical representation which is used) but most of the essential mechanisms can be retained, more realistic patterns can be derived.

Christaller (1933) presents quite a long analysis of dynamical questions. The ideas, however, are mostly qualitative and based on comparative static notions. The problem with a rigid structure such as those of the standard models is that it is difficult to see how new structures can evolve as parameters or other variables change. It is clear that one of the main objectives in the development of the SIA models presented in this chapter is to address the question of dynamics and evolution of such systems. We argue below, therefore, that another advantage of the representation used is that progress can be made in this direction which has been inhibited by the rigidities and technical problems of central place theory.

The next step therefore is to present a model obtained from an integration of the ideas offered earlier in this chapter and in Chapter 5 and to show how this can represent most of the ideas of central place theory. We should emphasise, of course, that this is an illustrative, methodological argument and if it is accepted, then much more work remains to be done.

6.4.2 An alternative model representation for central place theory

The models we have described earlier in this chapter are based on a now-substantial history of modelling methods involving spatial interaction and the location of activities in relation to discrete zoning systems. They have been referred to above and elsewhere, in the context of their relevance to central place theory, as SIA (spatial interaction-activity) models (Wilson, 1977-A). In the preceding sections of this chapter we have seen how these models can be extended to include structural variables endogenously, such as facilities for providing goods and services ('retail centres') and housing stock. It is this extension which allows the connections to be made between the SIA modelling style and central place theory. Variables like W_j^g measure the attractiveness of j for the provision of good (or service) g, and, as a composite attractiveness factor, we have seen that this would include measures of 'size' of facility, such as floorspace. Indeed, for convenience, we usually take that as the main measure.

The market area of a particular good will be determined by the size of the parameters like β^g in the spatial interaction model of consumers' response to the pattern of facilities $\{W_j^g\}$: if β^g is small, the market area is large, and vice versa. Note that, now, however, there is no restriction asserting that market areas cannot overlap. Typically, there will be overlaps which may arise due to market imperfections, lack of perfect information, and the fact that any particular 'g' covers a variety of goods and services within it (for which tastes and perhaps even prices will differ). Only when parameters like β^g tend to infinity will there be non-overlapping market areas, because then consumers will travel to the nearest centre which supplies g.

One of the main advantages of the SIA model system is that it is no longer necessary to make restrictive assumptions about the distribution of population: in the aggregate shopping models of Section 5.2 for example, zonal expenditures and populations were taken as e_i and P_i respectively. The former can be used to distinguish population in relation to spending power, the latter by size and density. Nor is it necessary to make simple assumptions about ease of travel in different parts of the city: the matrix $\{c_{ij}\}$ can be made to contain a rich amount of information about density of networks and, by adding an appropriate traffic assignment model, it can be made to reflect levels of congestion which vary spatially across the city.

6.4.3 An example of an S. I. A. model to be used as the basis for intra-urban central place theory

In this section, we assemble a complete model by integrating a number of the ideas which have been presented in earlier sections of this chapter. We give the model a Lowry-like structure which is shown in Figure 6.1. This assumes that households locate themselves around workplaces, and so the employment distribution tops the diagram followed by residential location and housing supply. In the original Lowry model, it is assumed that housing supply 'follows' population demand, so

Chapter 6

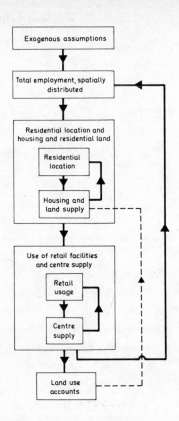

Figure 6.1 The structure of the Lowry model

that one of the extensions here is to deal with housing supply
more interestingly, and also with supply of residential land.
The spatially-distributed people generated by this model then
demand goods and services. This leads to the further extension
which models the supply of retail structures. This then
generates a substantial contribution to employment and so an
iterative-loop connection is shown to the first submodel. (The
way in which this iterative loop is solved has sometimes been
interpreted as a dynamic growth mechanism in itself -
cf. Batty, 1976, for example - and this should be carefully
distinguished from the other aspects of dynamics discussed here.)
The transport inputs are shown as exogenous (through the

c_{ij}-term), though (as noted briefly earlier) this could easily be extended to another kind of iterative connection to include the effects of congestion in the model.

In the model assembled below, we have attempted to choose levels of disaggregation and forms of composite attractiveness factors which are the simplest to exhibit all the main phenomena we would wish to incorporate into an SIA version of a central place theory model. A fully realistic model would, however, need much more detailed work and added complexity.

Total employment. The total number of jobs in j of type w:

$$E_j^w = \sum_g y^{wg} E_j^g + y^{wB} E_j^B \qquad (6.45)$$

(probably taking w to distinguish by income).[2] E_j^g and E_j^B are the numbers of jobs in retail sector g ($g = 1,2, \ldots, G$, say) and the basic sector, B, respectively, and the y^{wg} and y^{wB} are coefficients which break these down by w-categories. The E_j^g are supplied from the retail model below.

Residential location. We continue to take w as income group and now add k as house type to the usual locational variables. The Equation (5.26) or Section 5.2.5 can be modified to take the form

$$T_{ij}^{kw} = B_j^w E_j^w W_i^{reskw} e^{-\mu^w c_{ij}} \qquad (6.46)$$

where

$$B_j^w = 1/\sum_{ik} W_i^{reskw} e^{-\mu^w c_{ij}} \qquad (6.47)$$

Since this is a singly-constrained model, the distribution of population by type is

$$P_i^w = \sum_{jk} T_{ij}^{kw} \qquad (6.48)$$

As a form of composite attractiveness factor we can return to Section 5.4.4 and Equation (5.98) and disaggregate it to obtain, say

$$W_i^{reskw} = \prod_k (X_{1i}^k)^{\gamma_1^{wk}} \cdot X_{2i}^{\gamma_2^w} X_{3i}^{\gamma_3^w} X_{4i}^{\gamma_4^w} X_{5i}^{\gamma_5^w} \qquad (6.49)$$

Chapter 6

For simplicity, we only extend the X_{1i}-variable to take account of the house type disaggregation and this explains the product term in Equation (6.49). The full definitions can now be taken as

$$X_{1i}^k = H_i^k \qquad (6.50)$$

$$X_{2i} = L_i^{res} \qquad (6.51)$$

$$X_{3i} = (\sum_k H_i^k)/L_i^{res} \qquad (6.52)$$

$$X_{4i} = (\sum_k H_i^k)L_i^{tot} \qquad (6.53)$$

and

$$X_{5i} = \sum_{jg} W_j^g e^{-\beta^{1g}c_{ij}} \qquad (6.54)$$

where β^{1g} is defined in relation to the service model below.

The residential supply side: housing and land. We saw from the argument in Section 5.4.4 that the supply-side differential equations should be built in terms of housing and land variables, which are now H_i^k and L_i^{res}. We can extend Equations (5.104) and (5.105) to take the form

$$\dot{H}_i^k = \rho^k (\sum_{jw} \frac{h^w}{\lambda^w} T_{ij}^{kw} - q_i^k H_i^k) \qquad (6.55)$$

and

$$\dot{L}_i^{res} = \sigma(\sum_{jkw} \ell^k T_{ij}^{kw} - r_i L_i^{res}) \qquad (6.56)$$

where the constants are as before except that they are now disaggregated by k or w as appropriate. We have now added zone labels, i, to the q and r terms to allow for a more detailed 'rent' theory to be incorporated into the model. The equilibrium conditions can be written

$$\sum_{jw} \frac{h^w}{\lambda^w} T_{ij}^{kw} = q_i^k H_i^k \qquad (6.57)$$

and

$$\sum_{jkw} \ell^k T_{ij}^{kw} = r_i L_i^{res} \qquad (6.58)$$

Use of retail services. In Section 5.2.5 we showed how the
demand for retail services is generated from both homes and
workplaces and we also showed how the model could be disaggregated. Introducing a slightly new but obvious notation to deal
with the two separate origins of service sector trips, we can
write the model as

$$S_{ij}^{pg} = A_i^{pg} e_i^{pg} Y_i^p \hat{W}_j^g e^{-\beta^{pg} c_{ij}} \qquad (6.59)$$

where

$$A_i^{pg} = 1/\sum_j \hat{W}_j^g e^{-\beta^{pg} c_{ij}} \qquad (6.60)$$

and

$$Y_i^1 = P_i \qquad (6.61)$$

together with

$$Y_i^2 = E_i \qquad (6.62)$$

The superscript p now distinguishes origin of trip, with p = 1
representing home and p = 2, work.   This formalism also has
the advantage that with suitable definitions of g it could be
extended to include the demands on service sectors by other
sectors (over and above the demands of their employees). P_i
and E_i can be obtained from Equations (6.48) and (6.45)
respectively:

$$P_i = \sum_w P_i^w \qquad (6.63)$$

$$E_i = \sum_w E_i^w \qquad (6.64)$$

and the 'e_i' terms can be made to reflect the income
composition of the origin populations. If $e^{(1)gw}$ and $e^{(2)gw}$
are suitable conversion coefficients measuring the demands for
goods and services, then

$$e_i^{1g} = (\sum_w e^{(1)gw} P_i^w)/\sum_w P_i^w \qquad (6.65)$$

$$e_i^{2g} = (\sum_w e^{(2)gw} E_i^w)/\sum_w E_i^w \qquad (6.66)$$

Chapter 6

The composite attractiveness functions can be taken as

$$W_j^g = (W_j^*)^{\alpha_1} \prod_{g'} (W_j^{g'})^{\alpha_2^{g'g}} \qquad (6.67)$$

which builds on the ideas of Equation (5.16) of Section 5.2.5 above. In practice, the array of coefficients $\alpha_2^{g'g}$ would have to be substantially reduced in dimension. Here, they are designed to be able to reflect the effect in the attractiveness function on the supply of facilities for any one good (g') on the attractiveness for any other (g).

The total revenue attracted to facility g in zone j can then be calculated as

$$D_j^g = \sum_{ip} S_{ij}^{pg} \qquad (6.68)$$

and total employment as, say

$$E_j^g = \phi^g D_j^g \qquad (6.69)$$

for suitable coefficients ϕ^g. In Lowry's version of such a model, at this stage he includes an equation which computes the amount of land used in retail activities, as part of his land use accounts. For completeness, therefore, we can add such an equation as

$$L_i^{Rg} = \nu^g E_i^g \qquad (6.70)$$

which gives total retail land use as

$$L_i^R = \sum_g L_i^{Rg} \qquad (6.71)$$

using an obvious notation.

Retail supply side dynamics. We can now follow the argument of Section 5.3.6, and Equation (5.47) in particular, and write

$$\dot{W}_j^g = \varepsilon^g (D_j^g - k_j^g W_j^g) \qquad (6.72)$$

with equilibrium conditions

$$D_j^g = k_j^g W_j^g \qquad (6.73)$$

As in the residential case, we have added a zone label j, as
well as g, to the k-coefficient.

Land use accounts.   If we take

$$L_i^B = \nu^B E_i^B \qquad (6.74)$$

as the land used by the basic sector and L_i^U as being the amount
of unusable land, then the land-use accounting equation becomes:

$$L_i^U + L_i^{res} + L_i^R + L_i^B = L_i^{tot} \qquad (6.75)$$

We have now constructed all the necessary equations and can
investigate their solution properties in both a static and
dynamic sense. At any point in time, the model can be solved
as the original Lowry model, but there are additional equations
to solve for some supply-side variables. If the model was being
run for a situation in which only basic employment was given,
then this provides the starting point for Equation (6.45).
However, typically the employment variables will take values
from some previous time period and will include some service
employment. The equations must be solved iteratively for each
point in time. The employment distributions from
Equation (6.45) are fed into the residential location model
given by Equations (6.46)-(6.54). These in turn have to be
solved in a sub-iteration with the supply-side equilibrium
conditions (6.57) and (6.58).

This procedure generates the population distribution P_i^w
which forms the input to the service demand model. Equations
(6.65) and (6.66) would be computed first to give e_i^{pg} and
Equations (6.61)-(6.64) provide the other inputs to
Equations (6.59) and (6.60) (with attractiveness factors given
by Equation (6.67). These have to be solved in another sub-
iteration with the supply equilibrium conditions (6.73) (and
this also involves Equation (6.68) as part of that sequence).
The total service employment, sector by sector, is obtained
from Equation (6.69) and this provides the new input to the
overall iteration - back to step 1 and Equation (6.45).

The land-use Equations (6.71) and (6.74) have not been used explicitly, but it may be possible, if desired, to employ them in the style adopted by Lowry and input these to the land accounting Equation (6.75) and use this to estimate the amount of land still available for residential development. This could be achieved by the addition of an appropriate mechanism to the residential supply model, either through the attractive- ness terms or through the r_i term.

6.4.4 Urban dynamics and the S. I. A. model of central place theory

Now that we have assembled the complete model and investigated the solution procedure for one point in time, we can see how the model functions over time. This is determined, in part, by the specification of the exogenous variables or parameters of the model. These are all listed for convenience in Table 6.1. Much of the urban theory which is currently missing from the models is implicit in the assumptions which have to be made about the behaviour of these parameters over time, and we will discuss likely hypotheses about their behaviour in the future in this light.

The parameters are listed in Table 6.1 in the order in which they appear in the model as presented. It is also con- venient to list them under the headings of the theoretical assumptions which have to be made to predict them, and this is done in Table 6.2.

The first set of assumptions relates in various ways to the implicit account of the economy which appears in the model. The 'basic' sector is defined in many ways, and these defini- tions are sources of great controversy. Perhaps the most common definition is that it is 'export' activity - for the city that is, not the nation. c_{ij} has also been added to the list at this point as an exogenous input. It is characterised in the table as being a representation of the supply of trans- port (though there is also an element of interaction with demand, through congestion, in determining its actual value). The

Table 6.1 Model parameters for SIA central place system

Total employment

y^{wg}, y^{wB} Relate jobs to income earned from them.

E_j^B Distribution of basic employment.

Residential location

μ^w Distance parameter related to average length of journey to work for w-people.

$\gamma_\ell^w, \ell = 1 - 5$ The equivalent of the 'α' parameters for the disaggregated model. Records consumer preferences for house type, by w and in relation to other house types nearby.

Housing and land supply

h^w, ℓ^k Average expenditure on housing and on residential land.

λ^w Average number of workers per household.

q_i^k, r_i Costs of supplying type k houses and residential land.

ρ^k, σ Rates of return to equilibrium after a disturbance.

Use of retail services

β^{pg} Distance decay parameter by origin type and good or service.

$e_i^{pg}(e^{(1)gw}, e^{(2)gw})$ Expenditures (or other measures of demand) by g by w.

$\alpha_1^1, \alpha_2^{g'g}$ Parameters governing consumer taste in relation to centre structure.

ϕ^g Number of employees per unit of revenue.

ν^g Amount of land used in supply by sector, per employee.

Retail centre supply

k_j^g Cost of supplying retail facilities.

ε^g Rate of return to equilibrium after a disturbance.

Chapter 6

Table 6.2 Model parameters and assumptions associated with them

---

The economy (T6.21)

E_j^B	The 'basic' economy, assumed here to be one of the key driving variables.
c_{ij}	The supply of transport.
y^{wg}, y^{wB}	Implicitly gives the amount of wage income derived from jobs.
e^{ipg}, $(e^{(i)gw}, e^{(2)gw})$	The amounts available from household income to spend on goods and services.
h^w, ℓ^k	The amounts available from household income to finance residential services.
ϕ^g, ν^g	Elements of the service sectors' production function.

Economic/social characteristics (T6.2.2)

λ^w	The activity rate, which is partly a function of demand for employment, partly of demand for labour.
μ^w, β^{pg}	The transport deterrence coefficients, partly determined by consumer tastes, partly by transport and land use supply.
q_i^k, r_i, k_j^g	The costs of supply housing, residential land and retail structures. The production functions of these sectors.

Consumer behaviour (T6.2.3)

γ_ℓ^w, $\ell = 1,5, \alpha_1^1, \alpha_2^{g'g}$	Housing preference weights (in effect) and ditto for retail centres.

Relative dynamics (T6.2.4)

ρ^k, σ, ε^g	Related to the nature of the processes involved in change and in recovery after a disturbance.

---

y-coefficients are essentially a description of the size and
distribution of household income. Typically, in an advanced
economy, this has grown over a long period, but with a relative
distribution which has not changed very much. The proportions
of this spent on goods and services are then defined by the
e-coefficients. These are estimates of demand.

If the economy continues to grow, then we could expect
increasing demand for luxury goods and services. But in some
countries, it is argued that the limits to growth have been
achieved (Broadbent, 1977, argues this for example). (h^w, ℓ^k)
represent the same kinds of things for residential services.
The coefficients ϕ^g and ν^g are crude representations of the production functions of the service sectors: being the amounts of
employment and land needed for a unit level of activity in
relation to money units of revenues. Both sets of coefficients
can be expected to decrease over time in established sectors,
as capital replaces labour with technological advance (or to
maintain the rate of profit as some schools would argue). Even
for new sectors, growth and development is likely to be given
by increasing e-coefficients rather than ϕ^g or ν^g ones.

The economy, with all its interdependencies, is obviously
more complicated than is implied by the set of coefficients
which are discussed above. It remains an important research
task to extend the underpinning economic theory in urban models
of this type. Since interdependencies are represented by input-
output models, this provides one line of inquiry - and this is
one where some progress has been made, for example in seeing the
Lowry model as being a special case of an input-output model and
then extending it to incorporate a full set of inter-industry
relations (Macgill, 1977).

The three terms q_i^k, r_i and k_j^g are all included under the
heading of the economy. As the costs of supplying residential
or service facilities, they are composite representations of the
production functions of these sectors. These parameters are the
basis of much of the interesting dynamic behaviour outlined

earlier in the chapter, involving jumps for example, and can be
seen as related to thresholds.   But it is important to emphasise
that there are large elements of theory missing from the presentation so far. Beaumont and Clarke (1979) for example, argue
that these thresholds can be generated from a more general
knowledge of bid rent (or preference) functions on the one hand
and production functions on the other. More generally, these
terms form the subject matter of the 'new' urban economics
(*cf.* Richardson, 1977). We have shown them here as having a
spatial variation, and this in part, is to enable us to build
the spatial variation in rents predicted by economic theory.
In particular, we should note that if the theory can be extended,
these parameters would become functions of many other variables
which are already endogenous to the model system (eg. density
terms). This would, of course, imply the existence of a number
of new bifurcation properties.

The next set of parameters are grouped in Table 6.2 under
the heading of social and economic characteristics. They
summarise the outcome of economic and social processes which
interact. λ^W is an activity rate which is an outcome of the
demand for employment by members of households - say to obtain
income which is then used for a variety of purposes - relative
to the supply of jobs. This partly depends on taste, partly on
the attractions the economy has to offer. Because so many
factors are involved, it is difficult to make predictions for
the future except to say that the current trends towards auto-
mation in both manufacturing and service sectors suggest that it
is likely to decrease.   The two transport deterrence coefficients are also included under this heading because they in part
represent consumer preferences for travel - particularly in
relation to length of trip - but these are also partly constrained by the alternatives which are on offer to the consumer.
They will take lower values in a low density city for example.

Consumer preferences for housing and services (except in
relation to transport and accessibility issues, which are mostly

picked up in other parts of the model) are represented by the
exponents of the different factors in the composite attractiveness factors in the residential location and use of retail
facilities' models. Once again, it is difficult to predict
their variation over time.

Finally, the relative dynamics of elements of the system
are described by the coefficients ρ^k, σ and ε^g. We have
relatively little experience of the measurement of these and
this makes it even more difficult to make predictions.

This analysis shows that in order for the SIA model to be
used in dynamical analysis, general (often alternative) assumptions about the future have to be made - for example about
economic development - and these have to be translated into
assumptions about the exogenous variables and parameters of the
model. Although the list of parameters is a long one, so too
is the list of endogenous variables; the model, in principle,
still has a lot to tell us about the possible urban structures
which can emerge from, or which are compatible with, various
alternative assumptions about the future. In the next subsection,
we carry these ideas further and see what the model has to offer
in dynamical analysis.

6.4.5 The evolution of urban structure

In Section 5.4.3, we discussed in detail the problem of
modelling the evolution of urban structure for the retail sector.
Although we developed models for the other sectors, we have not
yet taken the discussion of evolution further in detail for a
system in which all the major submodels are fully coupled. This
is an appropriate place to begin to do so, because we can use
the example of the model of Section 6.4.3 and show how, in
effect, we are making a contribution to dynamic central place
theory. In Section 6.4.3, in discussing the solution of the
model equation system at a point in time, we in effect showed
how to use the model to get a static picture. This should produce patterns which have at least a family resemblance to the

classic ones of central place theory (and the equivalence is
likely to be closer as the transport deterrence factors become
very large - *cf*. Evans, 1973 and Wilson and Senior, 1974). We
then saw in the preceding section that to use the model in a
dynamic mode involves making assumptions about all the exogenous
variables and parameters. We now explore how to use the model
in dynamical analysis and then discuss in more detail the kinds
of assumptions which are needed about parameters.

Essentially there are three modes of operation for the
model in dynamical analysis: (1) historical analysis;
(2) conditional forecasting; and (3) as part of a planning
control system. All three are closely related in a number of
ways, as we will see. Historical analysis would involve
specifying the time series of all the parameters exactly (or as
near as can be obtained from data) and then using the model to
explain the evolution of urban structure which this implied.
The model's account could then be compared with reality.

Conditional forecasting is a very similar mode of operation,
to the above but the values of the parameters have to be
specified for a future time. The values predicted by the model
are conditional on these assumptions. This is closely related
to the third mode - the use of the model as part of a planning
system in the sense that the future settings of some of the
parameters may be elements of a plan: a run of the model then
represents the impact of the plan and the outputs of the model
can form inputs to various evaluation indices. To use the
model in a planning mode, the planning system of which it then
becomes part has to be specified. (Conversely, the model in an
historical or conditional forecasting mode needs assumptions to
be made about the effect of the planning system.) In the
example just used, this would involve noting the variables
which the planners could control and then using the model to
predict the impact of these. This is a well-defined and
commonly used role for models. Possible extensions of this role
were explored in Section 6.3 when these concepts were embedded
in a control theory framework of a formal kind.

There are also possibly more subtle uses in planning. For example, we have been much concerned with the stability of particular zones - whether development of a certain kind is possible or not (the DP and NDP states). This information may also be directly usable as part of the planning process: the goal would then be to adjust some of the controllable variables to maintain (or get rid of) a stability condition in a certain zone. Such ideas can be related to those of 'robustness' or 'resilience'. These different modes of operation have to be borne in mind in the explorations which follow. We will concentrate on historical analysis and conditional forecasting, but it will be easy to bear in mind the ways in which these ideas could be extended to the various planning modes.

First, let us consider an historical analysis for a particular town, beginning by returning to Table 6.2. Suppose we are considering a period in which a city has grown and we wish to explore the kind of urban structure which is likely to have evolved on the basis of the Section 6.4.3 model. The economy would have been growing, due to technological change, the multiplier effects of increase spending power, and capital accumulation. There are many alternative economic models of this process. However, in terms of Section T6.2.1 of Table 6.2, we can argue as follows. There would be a period when the main source of economic growth was in the basic sector and this would be located, often, in relation to resource availability rather than markets. The simple implicit demographic model implies a corresponding growth in population, proportional to both basic and service jobs. The main surge of such growth was during the Industrial Revolution. In later times, this sector will still be large in absolute terms but will have shrunk in relative terms. Much of it will have outgrown its original sites and started to look for new locations, inevitably less central because of the availability of cheap land. c_{ij} will have decreased consistently over a long period due to technological innovation and due to increased investment in transport both by

Chapter 6

the state, by firms and by individuals. In the context of
growth, the income and expenditure items in the T6.2.1 list
will, of course, all grow. There may be contrary trends,
particularly in recent times in the coefficients which represent the service sector production functions, ϕ^g and ν^g,
because of productivity improvements.

It is more difficult to be specific about many of the
other coefficients. The household activity rate has probably
decreased over time through younger members of families staying
in education longer and more women working as housewives. This
may now begin to increase again with an increasing tendency for
women to work. The transport deterrence parameters have consistently decreased over time: travel has become easier. But
again this may cease to be the case in the light of successive
oil crises. Typically, the costs of supplying a unit of urban
structure - the coefficients q_i^k and k_j^g - have probably decreased
over time, at least in real terms while the cost of land r_i,
will have increased due to its increased productivity.

The parameters of consumer behaviour are more difficult to
assess. One might conjecture that they have increased over
time as people have become more interested in variety and
quality of goods and services rather than in simply the supply
of necessities. This will have been reinforced by easier
travel, particularly by car, which opens up a wider range of
opportunities for comparison.

As noted earlier, relatively little is known about the
parameters ρ^k, σ and ε^g. Perhaps residential structure has more
inertia than that of service sectors, so that

$$\varepsilon^g \gg \rho^k \tag{6.76}$$

What does all this mean, then, for the workings of the
model? First, because we have used a simple implicit demographic assumption as noted above, the population will grow -
through growth in E_j^B, through growth in income and expenditure
coefficients (because these create service sector jobs) and

194

through various multiplier effects. This will produce a lot of increases which are like $e_i P_i$ increases, in our analysis of shopping model dynamics: that is, many more zones will progress from NDP to DP states with respect to both residential development of different types and retail structures.

Because of both land availability and other elements of the residential attractiveness function, this means that the city will spread out at increasingly lower densities and with increasing social polarisation. The richer people in the sub-urbs will then attract more service centres away from the city centre (which would be the most accessible and likely spot for the first one). As the economy grows also, thresholds would be crossed for the provision of higher order goods and a more differentiated hiearchy in the service sector would emerge. In other words, a reasonable picture of urban development is implied by the model which has been presented. The details could only be worked out in trials of numerical simulations for different particular settings of the parameter values. It would also be useful to incorporate fluctuations. Indeed, if the model was being run in historical mode, it would be possible to build in exogenously any particular historical accidents - like the development of particular centres which may run counter to the trends predicted by the model. Such events may force the model to select solutions from among the many where some of the DP-zones remain at zero levels of development in at least some respect and may distort the hierarchical pattern.

The most interesting feature of this kind of analysis would be to find the times and parameter values at which major struc-trual changes take place: the creation of new orders of centre, the onset of suburbanisation, the acceleration of social polari-sation and the interactions between these. This would be particularly interesting in a conditional forecasting mode. For example, is there a critical level of oil prices at which the suburban-rich/inner-city poor structures of our cities as we now know them would begin to change?

6.4.6 Comparisons with traditional central place theory

The comparison of the Section 6.4.3 model in its different modes of use with that of central place theory has been implicit throughout the discussion. A number of points can be drawn together here. The micro-level underpinning of the theory appears here as the spatial-interaction models of consumer behaviour together with the threshold parameters which appear in the supply-side differential equations. These generate structures, $\{W_j^g\}$, which although they overlap, are well defined by the spatial interaction parts of the model. A major addition is the incorporation of a much more mobile population, and the addition of associated structures as residential land and housing, than is evident in most statements of central place theory. The structures which develop do not have the rigidity of the central place theory ones and these, together with the mechanisms which have been added, make it much easier to build an evolutionary model. This flexibility may also make the model more useful in a planning context than has been the case in the past with central place theory. This is not to say that there are not severe remaining problems, which will be discussed in the final section, particularly that of modelling morphogenesis.

6.5 Some possibilities for further research

6.5.1 Introduction : development or evolution ?

It can be argued that there are two interesting kinds of issues involved in studying the dynamics of system structure - in this case urban structure. First, there is the development question: how does the system grow to take the form which it does? Then there is the question of evolution: how do new *kinds* of structure evolve? These issues are illustrated by the differences between developmental biology and the biology of evolution: how does an organism, examples of which we have seen and studied, grow? The evolutionary question is: how do new species emerge? Most of the techniques offered in this chapter relate to what are mainly developmental questions. On the other hand, to discuss even the possibility of evolutionary questions, we would have to know how to recognise a new 'species' in a city. There are probably evolutionary developments taking place currently: the trend, for example, towards the development of larger and larger organisations may provide an example. A more clear-cut example is the emergence of new industry, perhaps microelectronics or, more speculatively, an oil-free transport system.

The essential problem of *modelling* evolution involves knowledge, to be preprogrammed in some way, of what it is that can be produced. In an urban economic model, for example, there may be some function which is not currently performed and we might wish to predict the conditions under which it could emerge. This can only be done straightforwardly *if we know the production function in advance*. Otherwise, some sort of learning model is required and the construction of this is a much more difficult task.

In the next subsection, we discuss some of the possibilities for research on developmental problems and then we present an example to illustrate the above remarks on evolution. Some concluding comments are made in Section 6.5.4.

6.5.2 Further research on models in the development mode

The position which has been reached to date is that considerable theoretical insight has been achieved on the possible modes and mechanisms of change in cities and we can convince ourselves that these models contain a plausible account of what has happened in the past. However, we need more numerical experiments on the one hand and some real data to reinforce these on the other. In the former case, we stand to gain much insight on possible forms of development, and also in the case where the development is deterministic, but with different possible parameter values and initial conditions. If fluctuations are added to this, then this provides a methods for sampling the high-dimensional combinatorial space of possible equilibrium states which may exist at any one time and which may

generate many possible alternative development trajectories for
the whole system. The second style of work, seeking real data,
is likely to need much effort and ingenuity, since the data
required will not be easily available from standard sources.

It will also be possible to experiment with alternative
forms of model. In this chapter, we have concentrated on particular styles and these could be amended or changed fundamen-
tally. Any model with non-linearities and interdependence could
have interesting bifurcation properties. For example, in the
mathematical programming forms of the models used here, presented in Appendix 1 we have concentrated on consumers' surplus
maximisation as the driving force and the main element in the
objective function. Other elements could be considered. For
example, in Wilson (1978), there are terms which represent the
interests of consumer, producer and government and the parameters
which are the relative weights of these terms could themselves
have interesting critical points. It is thought that
Hotelling's (1929) famous example of the stability of the
locations of two ice-cream men on a linear beach may provide a
simple illustration of this idea.

We also saw that to use the models fully in a dynamic
sense we should try to get beyond comparative statics, and in
the preceding section we discussed this issue in terms of pre-
dicting the variation over time of all the exogenous parameters.
Another research question, therefore, is to approach this issue
more formally and to recall Zeeman's six steps, discussed in
Chapter 3: what we would then be trying to do is model the
trajectories of the parameters - adding differential equations
to represent their behaviour. At least the models present here,
together with the notion of the exogenous parameters as driving
variables, provide an element in a research agenda: this
element shows where the thrust of further research has to come,
which variables to attack next - to seek differential equations
for, or to embed into some planning framework.

6.5.3 An example of the evolution of new structures

We follow the argument in Wilson (1978) and consider a number of types of retail centre in a problem defined as follows. Let W_j be a vector of centre 'sizes' as usual, but now let it range over types - like corner shops or supermarkets - rather than zones. Let $f(W_j)$ be the cost of the jth type of facility and $F(W_j)$ be its capacity. Suppose a number of constraints are set so that a minimum number of facilities N, has to be available together with a minimum total capacity, C. Suppose there are n_j facilities in level j, and suppose that n_j is determined by minimising costs subject to the constraints:

$$\underset{\{n_j\}}{\text{Min}} \; Z = \sum_j n_j f(W_j) \tag{6.77}$$

such that

$$\sum_j n_j > N \tag{6.78}$$

and

$$\sum_j n_j F(W_j) > C \tag{6.79}$$

The parameters of this model are N and C. New orders of facilities would emerge if C was increasing over time and there were scale economies. Figure 6.2 contains a speculative graphical representation of the functions f and F against W. For small C, a large number of small facilities will minimise cost, while as C increases critical values will be reached at which one higher level facilities achieves cost minimisation, and so on.

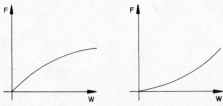

Figure 6.2 Capacity and cost functions vs. retail centre size

This can be made clear in an example where there are two types of facility and f and F are represented by linear functions:

$$f(W_j) = \alpha_j , \quad j = 1,2 \qquad (6.80)$$

$$F(W_j) = \beta_j , \quad j = 1,2 \qquad (6.81)$$

The problem can then be formulated as

$$\min_{\{n_1,n_2\}} \quad Z = \alpha_1 n_1 + \alpha_2 n_2 \qquad (6.82)$$

such that

$$n_1 + n_2 > N \qquad (6.83)$$

$$\beta_1 n_1 + \beta_2 n_2 > C \qquad (6.84)$$

and the non-negativity constraints:

$$n_1, n_2 > 0 \qquad (6.85)$$

This, of course, is a simple linear programming problem. To carry the illustration forward, suppose that

$$\beta_2 \gg \beta_1 \qquad (6.86)$$

so that $j = 2$ is the higher order facility. α_2 will then be greater than α_1, but if there are to be scale economies, we must have

$$\frac{\beta_2}{\beta_1} > \frac{\alpha_2}{\alpha_1} \qquad (6.87)$$

Then the various lines can be plotted as on Figure 6.3. The dashed lines represent

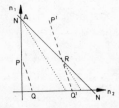

**Figure 6.3 Graphical representation of linear programming
model of evolution of shopping centre size**

$$\beta_1 n_1 + \beta_2 n_2 = C \qquad\qquad (6.88)$$

which allows C to vary as a parameter and the dotted lines
represent

$$\alpha_1 n_1 + \alpha_2 n_2 = Z \qquad\qquad (6.89)$$

for varying Z. In order to explore the solution to the linear
programming problem graphically, Z is varied until it reaches a
minimum, and we can then show the changing nature of the
solution with a parameter by varying C.

When C is such that (6.88) is the line PQ on the figure, it
is a redundant constraint and the optimum point is at A, with
$n_1 = N$ and $n_2 = 0$. However, when C increases so that PQ passes
through A, say to P'Q', then the optimum point becomes R, with
$n_2 > 0$. Thus a new order of facility has emerged. In this case
the production functions are presented by the coefficients
(α_j, β_j), $j = 1,2$, or more generally by the functions f and F.
The problem is that we will not usually know the production
functions of a new species. Indeed, if we do, it could almost
be argued that the problem is a developmental one, and only if
we have some surprise about the production function of the new
level is the process evolutionary: but this is a difficult, if
interesting, semantic point.

6.5.4 Concluding comments

We started this section by observing the distinction
between developmental and evolutionary dynamic and we have seen
that a major research programme is needed in each field. The
tools are available for modelling developmental problems, but
we have argued, particularly in relation to the simple example
of the previous subsection, that they are not yet available to
study evolution. This may turn out to be a problem of representation again: if a 'species' or organisation can be represented
by a set of characteristics, and an 'ecological' niche by a set
of attributes in relation to characteristics of systems elements
which can live in it, then this may provide the basis for

Chapter 6

evolutionary modelling. Even here, however, there are problems, since certain kinds of evolution in cities will modify the nature of potential niches.

Notes

1. There are exceptions to this broad argument: Angel and Hyman (1976), among others, for example, have shown how to build an SIA-type model with a continuous space backcloth.

2. We can also usefully recall the concept of 'subscript lists' (Wilson, 1974): a single symbol like w can, if required, be used in this formation to represent a vector of characteristics.

CHAPTER 7

MICRO-SCALE APPLICATIONS

7.1 Introduction

The micro-scale plays a fundamental role in geographical
(and other) theory. It is the level of theory concerned with
the behaviour of the basic elements of the system of interest -
usually people or organisations. At this scale, therefore, we
will usually be concerned with individual decision-making of
some kind or other. At present, only a relatively limited
range of examples is available. Most of the examples below
arise in transport modelling, and most of those from the study
of modal choice. However, it is easy to see in principle how
such examples generalise, and to aid the reader in adopting a
broader perspective, we begin with a brief review of broader
aspects of the problem.

In Section 7.3, we return to the application of catastrophe
theory to modal choice (we began with one of the initial illustrations in Chapter 1 - Section 1.2.3) and then, in Section 7.4,
examine mode choice in the context of bifurcation in the solution of differential equations. The third example, in
Section 7.5, is concerned with behaviour aspects of the speed-
flow relationship which arises in traffic engineering but which
also has an important impact on models at meso-scales through
the congestion relationships which are incorporated, explicitly
or implicitly, into $\{c_{ij}\}$ variables.

7.2 Some general considerations

At the micro-scale, the main state variable will, almost
by definition, refer to some aspect of individual behaviour.
The parameters, or control variables, are then likely to
characterise the environment within which the decision is made.

Chapter 7

Many micro-scale processes involve, *a priori*, characteristics which could generate interesting behaviour. Many models, for example, involve assumptions about an underlying optimisation process - such as utility or profit-maximising. If the objective function involves a non-linear relationship between state variables and parameters, many of the considerations of catastrophe theory are likely to apply. If a dynamic model is constructed, involving differential equations, then bifurcation can arise in the usual way when interdependence and non-linearities are present. The interdependence could arise from interactions between individuals or an interaction between a single individual and an environment.

Many behavioural models are expressed in terms of the probability of a state variable taking a value which represents a particular decision. The empirical nature of such probability distributions can, as Zeeman (1977-A, Chapter 1) points out, give a useful clue to the possible relevance of catastrophe theory, especially if an underlying optimising process can be identified. The most common probability distribution for the likelihood of a particular choice being made is the normal distribution, or some similar uni-modal distribution. This is consistent with there being a single 'most-likely' solution to the underlying optimising process. If, however, the distribution is bimodal, then Zeeman argues that this should make us think of the possibility of a cusp catastrophe mechanism, and if trimodal, of the butterfly. Possible distributions which occur over a range of behaviour in a particular phenomenon are shown in Figure 7.1. It is not difficult to see that such distributions could arise from a cusp model if the state variable is taken as a measure of types of behaviour, as in Figure 7.2.

Cases (a) and (b) in Figure 7.1 are the single solution extremes, case (c) involves positive values of the splitting parameter, while case (d) arises from measuring frequencies when the parameters are in the conflict set where there are two

204

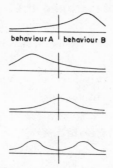

behaviour A ∣ behaviour B

Figure 7.1 Distributions of behaviour types

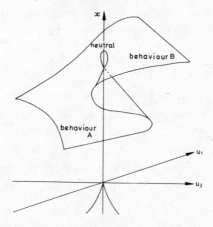

Figure 7.2 Cusp model for behaviour types

possible models of behaviour. The trimodal case links to the
butterfly because the latter involves another 'pocket' (surface)
of stable states between the upper and lower sheets of the cusp
surfaces.

In the examples which follow, we will draw attention to
these general features in the specific cases. In the concluding
sections, we speculate on how to apply them more widely.

Chapter 7

7.3 Hysteresis and modal choice

7.3.1 Hysteresis, catastrophe theory and modal choice

The theoretical ideas presented in this subsection first appeared in Wilson (1976-A) and were subsequently developed (Wilson and Kirkby, 1980) to revise the idea of the nature of the splitting factor in this particular model. Originally, the purpose of this model-building exercise was to teach the author something about catastrophe theory. It turns out, in this case, not so much to illustrate the notion of jumps, but that of hysteresis, which is potentially important in modal choice.

We begin by briefly recalling the argument of Section 1.2.3. The state variable, x, is taken as choice of transport mode, and the two control variables are u_1, representing habit, and u_2 taken as proportional to the difference between modal costs:

$$u_2 = k(c_2 - c_1) \qquad (7.1)$$

where c_1 and c_2 are the costs, for a particular origin-destination pair, for modes 1 and 2. The overall picture is presented in Figure 7.3 and a section of this figure, for $u_1 < 0$ so that habit factors are operating, is given as Figure 7.4. We assume that the perfect delay convention

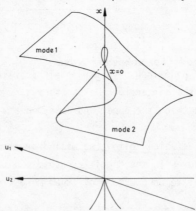

Figure 7.3 Modal choice and the cusp catastrophe

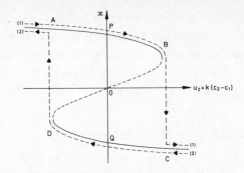

Figure 7.4 A section of the modal choice cusp surface

operates. Two trajectories are shown on Figure 7.4 representing:
(1) an increase in c_2 relative to c_1 and (2) a reverse increase
of c_1 relative to c_2.   Suppose the individual is initially on
the left-hand side of trajectory (1). At A, a new 'state'
becomes possible, but the jump to the lower surface does not
come until B. If there was no habit, or the perfect delay convention was operating, the jump would come at P. If the
situation is reversed, the jump back is not at C, but at D. The
path of the two trajectories traced out as the dashed curve
ABCD is an example of a classical hysteresis curve in its shape.

   The interesting empirical question is to seek to measure
the degree of habit (or any other relevant factors) which
causes varying degrees of hysteresis within a population. At
present, the model does no more than suggest a mechanism, but
empirical evidence in support of it has been produced - as
noted in Section 7.3.3 below, and this suggests that it would
be worthwhile to try to build such mechanisms into more conventional modal choice models. It is certainly important from a
policy point of view: the hysteresis effect implies that if
passengers (and, say, more than expected) are lost as a result
of a fare increase, a corresponding fare decrease will not be
sufficient to win them back again.

   This kind of mechanism could be considered to arise from a
utility maximising procedure and indeed Goodwin (1977) provides
such a model which is discussed in the next subsection. The

remaining general point to emphasise is that this example is
not concerned with jumps - by definition that actual x-value is
of no consequence beyond it being on the upper or lower sheet -
but it does show how to generate an appropriate hysteresis
effect in relation to control variables which make intuitive
sense.

7.3.2 A mechanism for hysteresis

Goodwin (1977) has presented a detailed mechanism for
showing how a habit factor can generate a hysteresis effect.
He builds up his argument effectively from simple to more complicated cases and also shows how to tackle the aggregation
problem: linking the individual model with a result for a
population as a whole. The main elements of the argument are
summarised below.

The simplest model is to assume that, for an individual,
he chooses the mode which is cheapest (in some 'generalised cost'
sense). Assume there exists a value E such that the probability
of choosing a mode, 1 say, is

$$p_1 = \begin{cases} 0 & G < E \\ 1 & G > E \end{cases} \tag{7.2}$$

where G represents modal cost differences. That is,

$$G = c_2 - c_1 \tag{7.3}$$

and E as possibly non-zero allows for a modal bias which has not
been included within G. This is a step function and is shown
graphically in Figure 7.5.

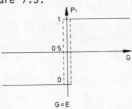

Figure 7.5 Modal probability vs. cost difference : step function

E can now be assumed to vary across the population.
Assuming that it is normally distributed with mean 0 and
variance 1, the proportion of the population choosing mode 1 on
this model is

$$p_1(G) = \frac{1}{2\pi} \int_{-\infty}^{G} e^{-E^2/2} \, dE \qquad (7.4)$$

and this generates an S-shaped curve as shown in Figure 7.6.

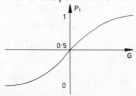

**Figure 7.6 Modal probability vs. cost difference : modal bias
normally distributed**

Habit is now introduced by means of a threshold, h.   The
model becomes

$$p_1 = \begin{cases} 0 & G > E + h \\ 1 & G < E - h \end{cases} \qquad (7.5)$$

with the possibility of the passenger using either mode in

$$E - h < G < E + h \qquad (7.6)$$

To get the result for the population as a whole, E is still
assumed to have a normal distribution, but there are two curves
according to whether G is increasing or decreasing:

$$p^{(1)}(G) = \frac{1}{2\pi} \int_{-\infty}^{G+h} e^{-E^2/2} \, dE \qquad (7.7)$$

$$p^{(2)}(G) = \frac{1}{2\pi} \int_{-\infty}^{G-h} e^{-E^2/2} \, dE \qquad (7.8)$$

which produce the lower and upper curves, respectively, of
Figure 7.7.   The original 'no-habit' curve is shown as dotted
between the other two.   When there is a reverse shift in G, as
at X, there must be a change of the order of 2h for any change
to be made at all.

Chapter 7

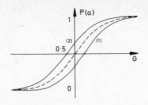

Figure 7.7 Modal choice probability for the whole population

The problem now is that the probability of an individual to move at any particular time in response to a particular cost change depends on the previous history of moves relative to cost changes. This leads to quite complicated possible hysteresis behaviour for the population as a whole, of which Goodwin explores examples. He also shows how to make the habit factor probabilistic in the population as a whole.

7.3.3 Empirical evidence

Blase (1979) has been able to use traffic-count data for London to test for the existence of a hysteresis effect. His data ranged over 48 months and included a period over which petrol prices rose substantially. He considered mainly weekend traffic on the basis that this was potentially more elastic. Some plots of his main data are shown in Figure 7.8. The traffic level against petrol prices, with the number of the months shown by each point, is shown in Figure 7.9. The different circled clusters relate to different time periods and he was able to fit a number of regression lines to represent three of the sections of a hysteresis cycle. The theoretical cycle is shown in Figure 7.10 and the estimated line, with confidence limits, in Figure 7.11.

7.4 Dynamic modal choice models and bifurcation

7.4.1 Some principles

This section is a report of the work of Deneubourg, de Palma and Kahn (1979). They consider an unlabelled origin-destination pair between which a number of trips, D, has to be distributed between two modes. To facilitate a consideration of extentions in the final subsection, we amend their

Figure 7.8 Petrol price and weekend traffic levels

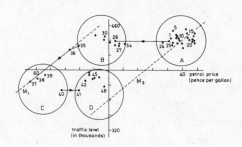

Figure 7.9 Saturday traffic vs. petrol price

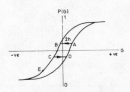

Figure 7.10 Modal choice hysteresis loop

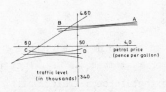

**Figure 7.11 Fitted lines for sections of hysteresis loop,
with confidence limits**

notation slightly and take x_k as the number of trips by mode k.
Thus

$$\sum_k x_k = x_1 + x_2 = D \tag{7.9}$$

D is considered, for most of their paper, as a parameter, and
we will consider only that role here. Dynamical equations for
x_k are then given by

$$\dot{x}_k = D_k - x_k \tag{7.10}$$

where

$$D_k = \frac{DA_k}{\sum_{k'} A_{k'}} \tag{7.11}$$

and so A_k is the attractiveness of mode k. In this sense, these
equations have a resemblance to the shopping centre structural
equations. Typically, the A_k's are each non-linear functions of
x_1 and x_2, and the non-linearity and interdependence generate
the bifurcation properties in the usual way. The equilibrium
conditions, extending the notation slightly to recognise this
functional dependence, are given by

$$\frac{DA_k(x_1,x_2)}{\sum_{k'} A_{k'}(x_1,x_2)} = x_k \tag{7.12}$$

In the next two subsections, we consider the consequences of
making specific alternative assumptions for the form of the A_k's.

7.4.2 Model 1: attractiveness proportional to speed

Assume

$$A_k = v_k^{p_k} \tag{7.13}$$

where v_k is the average speed of mode k and the p_k's are constants. It is assumed that there is no direct interaction
between the modes (except through the competition terms of the
denomination of the right hand side of (7.11)). If mode 1 is
taken as car and mode 2 as bus, it is assumed that car speeds
decrease with usage, because of congestion, while bus speeds

improve initially as supply responds to demand, and then decreases. These assumptions are expressed by

$$v_1 = \frac{1}{a + bx_1} \qquad (7.14)$$

$$v_2 = \frac{dx_2^n}{c + sx_2^r} \qquad (7.15)$$

for constants a, b, c, d, s, n and r. It is assumed that

$$n < r \qquad (7.16)$$

The example is developed for the simple case where

$$p_1 = p_2 = n = r = b = s = 1 \qquad (7.17)$$

so that

$$A_1 = v_1 = \frac{1}{a + x_1} \qquad (7.18)$$

$$A_2 = v_2 = \frac{dx_2}{c + x_2} \qquad (7.19)$$

Hence the Equation (7.10) can be written for each mode in turn as

$$x_1 = \left[\frac{D}{a + x_1} \right] \bigg/ \left[\frac{1}{a + x_1} + \frac{dx_2}{c + x_2} \right] - x_1 \qquad (7.20)$$

$$x_2 = \left[\frac{Ddx_2}{c + x_2} \right] \bigg/ \left[\frac{1}{a + x_1} + \frac{dx_2}{c + x_2} \right] - x_2 \qquad (7.21)$$

The equilibrium conditions are

$$\dot{x}_k = 0 , \quad k = 1,2 \qquad (7.22)$$

Setting the right hand side of (7.20) to zero gives, after some manipulation

$$(x_1 - D)(dx_1^2 + (ad + 1)x_1 - c - D) = 0 \qquad (7.23)$$

This involves substituting for x_2 using (7.9) as

$$x_2 = D - x_1 \qquad (7.24)$$

Equation (7.23) shows that there are three equilibrium values for x_1, and Equation (7.24) can be used to give the corresponding values for x_2. One is clearly

$$x_1^1 = D , \quad x_2^1 = 0 \tag{7.25}$$

Then, the other two solutions for x_1 are obtained by solving the quadratic equation part of (7.23):

$$x_1^+ = \frac{-(ad + 1) + [(ad + 1)^2 + 4(c + D)d]^{\frac{1}{2}}}{2d} > 0 \tag{7.26}$$

This notation is used because it can be seen from the form (7.26) of the coefficients that this solution if positive.

$$x_1^- = \frac{-(ad + 1) - [(ad + 1)^2 + 4(c + D)d]^{\frac{1}{2}}}{2d} < 0 \tag{7.27}$$

Because $x_1^- < 0$, this is not relevant and so is not considered any further. The x^+ solution given by (7.26) is acceptable only if

$$x_1^+ < D \tag{7.28}$$

which, of course, is equivalent to requiring

$$x_2^+ > 0 \tag{7.29}$$

This condition, again after some manipulation, can be expressed as

$$D > \frac{1}{2}\left[-a + (a^2 + \frac{4c}{d})\right]^{\frac{1}{2}} \tag{7.30}$$

Let D_c be the value of D at which equality holds in (7.30). Then, for $D > D_c$, x_1^+ is a possible solution.

The next step in the argument is to examine the stability of the solutions. It can easily be seen from a consideration of the signs of $\dot{x}_k$ in (7.10) for the different solutions that x_1^1 is stable when $D < D_c$ and not otherwise; and vice versa for x_1^+. Thus, if we plot x_1 equilibrium values against D (and x, using (7.24)), we obtain the plots of Figure 7.12. This is a familiar kind of bifurcation plot.

214

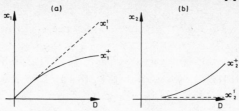

Figure 7.12 Modal choice bifurcation diagrams : (a) x_1 vs. D, (b) x_2 vs. D

7.4.3 Model 2 : addition of psychological factors

The attractiveness factors now take the form

$$A_k = v_k^{p_k} F_k \tag{7.31}$$

where the factors F_k are

$$F_k = \Theta_k + \alpha_k x_k \tag{7.32}$$

The Θ_k's are taken as measuring the effects of publicity and the α_k's, of imitation: the more people who use a mode, the more popular it becomes. Much simpler forms are taken for the velocity forms:

$$v_1 = \frac{1}{x_1} \tag{7.33}$$

$$v_2 = x_2 \tag{7.34}$$

and

$$p = q = 1 \tag{7.35}$$

Thus

$$A_1 = (\frac{\Theta_1}{x_1} + \alpha_1) \tag{7.36}$$

$$A_2 = \Theta_2 x_2 + \alpha_2 x_2^2 \tag{7.37}$$

and Equation (7.10) can be written for this case as

$$\dot{x}_1 = D(\frac{\Theta_1}{x_1} + \alpha_1)/(\frac{\Theta_1}{x_1} + \alpha_1 + \Theta_2 x_2 + \alpha_2 x_2^2) - x_1 \tag{7.38}$$

215

Chapter 7

$$\dot{x}_2 = D(\Theta_2 x_2 + \alpha_2 x_2^2)/(\frac{\Theta_1}{x_1} + \alpha_1 + \Theta_2 x_2 + \alpha_2 x_2) - x_2 \qquad (7.39)$$

All the parameters are taken as positive. The problem is further simplified by assuming that there is no publicity for the car. Then (7.36) shows that

$$A_1 = \alpha_1 \qquad (7.40)$$

and is constant and the equations simplify to

$$\dot{x}_1 = \alpha_1 D/(\alpha_1 + \Theta_2 x_2 + \alpha_2 x_2^2) - x_1 \qquad (7.41)$$

$$\dot{x}_2 = D(\Theta_2 x_2 + \alpha_2 x_2^2)/(\alpha_1 + \Theta_2 x_2 + \alpha_2 x_2^2) - x_2 \qquad (7.42)$$

Again, we find that

$$x_1^1 = D, \quad x_2^1 = 0 \qquad (7.43)$$

is a stationary state, and using (7.24), it can be shown that the condition for other stationary states is

$$\alpha_2 x_2^2 + (\Theta_2 - \alpha_2 D)x_2 + (\alpha_1 - D\Theta_2) = 0 \qquad (7.44)$$

This equation can have no, one or two physically acceptable solutions. In this analysis, Θ_2 and D are taken as the parameters of the model and the existence of different forms of solutions are related to them. The argument is a messy one and the reader is referred to the paper for the details. However, it is interesting, as in certain previous examples, that bifurcation behaviour can be obtained from simple assumptions and elementary mathematics - in this case a knowledge of the nature of the roots of quadratic equations.

The main results are summarised in three diagrams. In Figure 7.13, the different parts of the Θ_2-D plane are shown, according to the number of positive roots of Equation (7.44). When stability is analysed, the bifurcation diagrams in Figure 7.14 are obtained. These show x_2 plotted against D for two different Θ_2 values. The first of these is like Figure Figure 7.12(b), but the second, 7.14(b), shows a new kind of bifurcation behaviour. For $D_c < D < D_c^{(2)}$, there are two possible *stable* solutions for x_2, and the one chosen will depend on fluctuations.

216

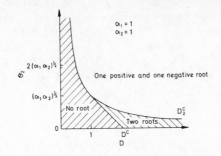

Figure 7.13 Conditions for solution of equations (7.44)

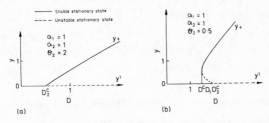

Figure 7.14 Bifurcation diagrams for the cases (a) $\theta_2 > (\alpha_1 \alpha_2)^{\frac{1}{2}}$, (b) $\theta_2 < (\alpha_1 \alpha_2)^{\frac{1}{2}}$

The two models presented above rely on specific assumptions about various parameter values and very much on the fact that only two modes are involved. More generally equations for model 1 would involve taking

$$A_k = \left[\frac{a_k x_k^{n_k}}{b_k + c_k x_k^{r_k}} \right]^{p_k} \tag{7.45}$$

using an obvious notation, and for any number of modes k. The second models could be incorporated if function factors F_k were added:

$$A_k = \left[\frac{a_k x_k^{n_k}}{b_k + c_k x_k^{r_k}} \right]^{p_k} F_k \tag{7.46}$$

217

Chapter 7

The differential equations would be (for the first case):

$$\dot{x}_k = D \left[\frac{a_k x_k^{n_k}}{b_k + c_k x_k^{r_k}} \right]^{p_k} \Big/ \sum_{k'} \left[\frac{a_{k'} x_{k'}^{n_{k'}}}{b_{k'} + c_{k'} x_{k'}^{r_{k'}}} \right]^{p_k} - x_k \tag{7.47}$$

with equilibrium conditions

$$D \left[\frac{a_k x_k^{n_k}}{b_k + c_k x_k^{r_k}} \right]^{p_k} = x_k \sum_{k'} \left[\frac{a_{k'} x_{k'}^{n_{k'}}}{b_{k'} + c_{k'} x_{k'}^{r_{k'}}} \right]^{p_k} \tag{7.48}$$

These are obviously very difficult to solve in general. Even with the restrictive condition $p_k = n_k = r_k = 1$, they are polynomials of order k (with $x_k = 0$ always a solution) (when there are k modes in all). The addition of F_k factors would normally increase the order of the polynomial by the order of the highest F_k expression.

We return briefly in Section 7.6 below to a more general consideration of this type of model.

7.5 The speed-flow relationship and the fold catastrophe

7.5.1 The empirical results

The time taken to travel from one point to another in a road network is an important feature of many urban models. Proper estimates of such a quantity involve taking account of the effects of congestion. The basis for doing this is to model the relationship, on a link of a road, between speed and the flow of vehicles on the link - the so-called speed-flow relationship. A third variable can be used to achieve the best insight into the form of the relationship: the density (or concentration) of the traffic on the link.

The empirical graphs connecting speed with density and flow with density for a link of a given type are shown in Figure 7.15. As one would expect, speed declines steadily with density, and the interesting feature of the second diagram is that flow is two-valued as a function of density. As density increases, flow increases up to a maximum. Beyond that, the reduction in speed reduces the flow until, in the limit, speed is zero and so is flow.

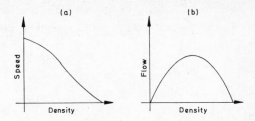

Figure 7.15 Traffic : speed–density and flow–density curves
These two graphs can be combined to give the speed-flow
relationship shown in Figure 7.16. This is also two-valued for
flow as a function of speed. Traffic engineers describe the
upper part of the curve as the zone of 'normal-conditions' and
the lower curve as the 'forced-conditions' zone (*cf.* O'Flaherty,
1974, p. 220). In normal conditions, density is sufficiently
low for drivers to be making their own judgements about speed;
in forced conditions, speed is determined by the vehicles in
front. The shaded area is sometimes called that 'unstable-
condition' zone - the situation can fluctuate between free and
forced states.

Figure 7.16 Traffic : speed–flow curve

7.5.2 A behavioural model and the fold catastrophe

The curve in Figure 7.16 is certainly folded. Dendrinos
(1978-A) offers a behavioural model which explains and generates
this fold. The essence of his argument turns on the construction of a utility function to describe driver behaviour. This
function has two components: first, a term which increases
with speed S (up to a bound S*), and secondly one which
decreases with flow - in relation to a given speed - because of
the increased possibility of accidents and loss of comfort.

219

Chapter 7

Let V be the flow and let C be the maximum flow (which traffic engineers sometimes call 'capacity'). Then the two components can be written

$$u^+ = u^+(S) \tag{7.49}$$

$$u^- = u^-(V/C,S) \tag{7.50}$$

and total utility is

$$u = u^+ + u^- \tag{7.51}$$

Since u^+ is supposed to be bounded, it takes the logistic form shown in Figure 7.17(a). u^- is assumed to decrease with increasing S, and more reversely for different V/C ratios, as shown in Figure 7.17(b). Thus, as V increases, the sum curve, u, changes shape as shown in Figure 7.18.

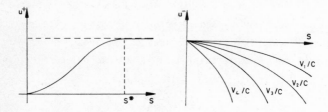

Figure 7.17 Driver utility components vs. speed and flow

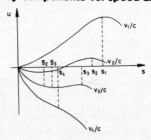

Figure 7.18 Total driver utility vs. speed

Figures 7.17 and 7.18 are immediately reminiscent of the mechanism used by Poston and Wilson (1977) in the argument about shopping centre size - described above in Chapter 4. And so it turns out. The driver choose the speed which maximises u. His 'environment' is given by the V/C ratio, and it is this which

we treat as the varying parameter for the problem. The speeds
chosen for V_1, V_2, V_3 and V_4 are plotted on Figure 7.18. V_4 is
choosen to represent capacity, $V_4/C = 1$, and u has a point of
inflexion there. The lower part of the fold curve is generated
by the local minima of u, and two examples are shown as $\bar{S}_2$ and
$\bar{S}_3$. If S is plotted against V/C, the fold curve of Figure 7.19
is obtained. The empirical fold curve is an aggregate of this,
averaged over a population of drivers.

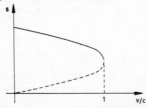

Figure 7.19 Speed–flow curve and the fold catastrophe

This is a neat and interesting argument. The only
possible weakness in it is that Dendrinos, as in another context in his work described in Chapter 4, assumes that the
unstable states are observable and describe the forced-driving
conditions. An alternative would be to make the direct assumption that, for density of traffic exceeding a critical point,
the driver essentially *has no choice of speed* and so the
utility function is flat, as shown for different V-values (for
densities above the critical level) in Figure 7.20.

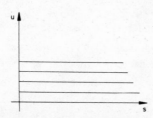

**Figure 7.20 Utility functions vs. speed at densities above a
critical value**

7.6 Concluding comments

In this chapter, four different mechanisms which are relevant to micro-scale models have been presented - ranging in examples from the very general to the very specific, though all the mechanisms have general applications. Zeeman warns us that bimodality or trimodality in probability distributions which turn up in behavioural work can indicate behaviour based on either the cusp or butterfly catastrophes. The first modal choice example is, in effect, an application of this: for certain cost differences, the probability distribution over a population would be bimodal and this is 'explained' using the cusp catastrophe with a habit variable as a splitting factor.

The third example possibly is the one which could generate the widest range of applications of the four. The application was to modal choice but we saw that the differential equations were similar in form to those we used for shopping centre evolution in Chapters 5 and 6. Indeed, the form of the equations could apply whenever choices involving attractiveness functions are used. What the example shows us, therefore, is how relatively simple assumptions can generate interesting bifurcation behaviour.

We saw that the fourth example was based on a fold-generating mechanism we had seen before, though the particular application is ingenious. We will see more examples of this arising in other disciplines in Chapter 8.

In the second and fourth examples, the models relate directly to individual behaviour, in each case with an 'environment' represented by variables which are the control variables in the catastrophe theory model. The third example is almost at the meso scale, but involves 'behavioural' hypotheses on the interactions between individuals and so has been included here. We might expect to see developments with each of these kinds of model. First, an interesting development would be a model which represented individual behaviour but with an environment generated by the explicit aggregation of all this

behaviour. Secondly, more micro-hypotheses might be added to
meso-scale representation so that bifurcation could be investigated in these. This might particularly apply to micro-
simulation models (*cf*. Clarke, Keys and Williams, 1979) whose
underlying meso-structures are accounting equations of the type
discussed in Section 5.5.3.

Chapter 8

Applications in other disciplines and some new results for urban systems

8.1 Introduction

In this chapter, we explore the progress of applications of catastrophe theory and bifurcation in a number of other disciplines and see if we can learn any more about possible analytical methods for urban systems. The most substantial progress has been in disciplines like physics, especially optics, and engineering (beams, stability of ships, and so on, switching problems in electrical and chemical engineering). However, these problems are unlike those we appear to face in urban analysis (though a possible exception is the study of phase transitions in physics, which is still a controversial matter). There is least progress in the social sciences, where we might expect to look for analogies. It is not easy to see useful analogues from the applications so far in psychology or sociology (except possibly in being aware of the consequences of bimodality or trimodality in behavioural models: that is, suggesting an underlying cusp or butterfly manifold).

So which disciplines are likely to be most useful for urban analysis? The basis of the answer to this question lies in Weaver's (1958) categorisation of different types of problems, or systems, in science: I - simple, II - disorganised complexity, and III - organised complexity. A type-I system is described by a very small number of variables, the other two by a large number. A disorganised system has relatively weak connections between its components and statistical averaging methods can be used for model building. An organised, type-III, system, has strong connections and no general methods are available for modelling. Urban systems are complex, in part disorganised (so that, for example, entropy maximising methods can be

225

applied - as in Wilson, 1970) and in part organised. So we need
to seek insights from disciplines which share these characteristics. Hence, in the rest of this chapter, we look to
physical chemistry, biology and ecology (Sections 8.2-8.4). We
find a general concern with interacting populations and find
some new methods whose application in urban analysis are discussed in Section 8.6 below. We also briefly review some
applications in economics in Section 8.5.

8.2 Physical chemistry

8.2.1 Kinetic equations: interacting mixtures

We saw in Chapter 2 that non-linearities together with
interactions between system components generate bifurcation
behaviour. In physical chemistry, this situation arises with
interacting chemicals in a mixture. Let x_i be the proportion
or amount of molecule i in a mixture, and then the dynamics of
the system is described by a set of equations of the form

$$\dot{x}_i = f_i(x_1, x_2, \ldots, x_n) \tag{8.1}$$

The interest arises because the function f_i can take a simple-
looking form, but one which turns out to generate interesting
bifurcation behaviour. Consider, for example, the following
autocatalytic reaction (Nicolis and Prigogine, 1977, p. 170):

$$A + 2X \underset{k_2}{\overset{k_1}{\rightleftarrows}} 3X \tag{8.2}$$

$$X \underset{k_4}{\overset{k_3}{\rightleftarrows}} B \tag{8.3}$$

The reaction rates are k_1-k_4. X is a catalyst to the reaction

$$A \rightleftarrows B \tag{8.4}$$

and in addition, through (8.2), catalyses its own production.
If the same letters are used to denote amounts of the chemicals,
and if the mixture is kept in contact with a reservoir of A and
B molecules in such a way that the proportions of those mixtures
are kept constant, then the differential equation for X is

$$\dot{X} = -k_2X^3 + k_1AX^2 - k_3X + k_4B \tag{8.5}$$

The contact with the reservoir means that the mixture is an open
system and is thus not in a state of *thermodynamic* equilibrium,
which is the other condition in thse particular problems for
generating bifurcation behaviour. Thus, typically, the differential equations (8.1) take forms like (8.5) with the functions
f_i as polynomials.

It is immediately clear from reference to Chapter 1 above
that X in Equation (8.5) will exhibit bifurcation behaviour
based on the cusp catastrophe: the right-hand side is a cubic
expression in X and the equilibrium states are obtained by
setting it to zero. This is a transformation of the cubic
which generates the cusp surface. A flavour of this is obtained
from Figure 8.1 (take from Nicolis and Prigogine, 1977). The
parameter b is given by

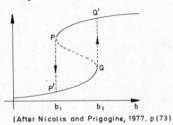

(After Nicolis and Prigogine, 1977. p(73)

Figure 8.1 Chemical concentration vs. ratio of reaction rates

$$b = \frac{\sqrt{3}}{g} k^{3/2} \tag{8.6}$$

where

$$k = k_3/k_2 \tag{8.7}$$

k_3 and k_2 are two of the reaction rates in (8.2) and (8.3). As
usual, the upper and lower parts of the curve are stable states
and the dashed middle part is unstable. The possibilities of
both jumps and hysteresis are present in the usual way. In
particular, as Nicolis and Prigogine note, such mechanisms form
the basis for modelling explosive reactions (Gray, 1974). The
basis of change in the 'control variable', b, obviously turns

227

on changes in the reaction rates k_2 and k_3.   These can alter with changes in the external conditions: for example, by changing the relative concentrations of A and B, or perhaps the temperature, T.

This example also illustrates the role of fluctuations in the models of the Brussels school.   The argument is as follows. Small fluctuations around stable states will usually simply lead to a return to the equilibrium state.   If, however, the system is sufficiently near to a critical point, say Q, or the fluctuations are large enough to reach the unstable states on PQ, then they can be amplified by the existence of the unstable states and lead to a jump to the other stable state *before* the critical point is reached.   This amplification is the source of the creation of 'order from fluctuations' which we will discuss further below.

A second example both further illustrates the principle of setting up kinetic equations and is interesting in the context of other examples in Chapter 2 and below.   Consider the reactions (again following Nicolis and Prigogine, 1977):

$$A + X \xrightarrow{k_1} 2X \tag{8.8}$$

$$X + Y \xrightarrow{k_2} 2Y \tag{8.9}$$

$$Y + B \xrightarrow{k_3} E + B \tag{8.10}$$

and again assume that a thermodynamic non-equilibrium state is maintained by keeping the concentrations of A and B constant. The differential equations for the proportions of X and Y are then

$$\dot{X} = k_1 AX - k_2 XY \tag{8.11}$$

$$\dot{Y} = k_2 XY - k_3 BY \tag{8.12}$$

which are our old friends the Lotka-Volterra prey-predator equations.   The molecule X is 'born' by autocatalysis by (8.8) and is 'eaten' by the predator Y in (8.9).   As we know, the

228

solution to such equations is a structurally unstable cycle. In
this case, a change in parameters implies a jump to a new orbit.

In general, sets of reactions can be identified which
exhibit the full range of solutions to simultaneous non-linear
differential equations and which have rich bifurcation properties. We consider one more example in the next section in the
context of explaining the ideas of 'dissipative structures' and
'order from fluctuations'.

8.2.2 Dissipative structures: order from fluctuations

In physical chemistry there is a concern both with
microscopic and macroscopic properties of systems. At the microscopic scale, the laws of thermodynamics tell us that energy is
distributed randomly across molecules: there is no 'structure'.
Prigogine and his colleagues have shown that, at the macroscopic scale (and in some cases, at what we have earlier called
the meso scale), structures can emerge provided that the system
is maintained in a state far from its thermodynamic equilibrium.
At thermodynamic equilibrium, the random distribution of energy
is dissipated among the molecules without the creation of
structure; far from equilibrium, there is the possibility of
macroscopic structure where, in effect, some of the energy is
used to maintain this structure: hence the notion of dissipative
structures.

These structures arise out of bifurcation properties, but
first it is useful to restress the distinction between thermodynamic equilibrium and steady state which is used here. The
former concept applies in thermodynamics to closed systems,
where there is no exchange of matter or energy with the environment. When the system is open, there is such an exchange. For
example, in the chemical mixtures used as examples in
Subsection 8.2.1, the maintenance of constant proportions of A
and B is a supply of matter. An alternative is to supply energy
in the form of heat. If this is done in a smooth way, the
systems will still achieve steady states - and the possible

Chapter 8

confusion with earlier discussions arises because these were called equilibrium states, for example in Chapter 5.

Bifurcation arises at critical points in the usual way and we saw one example earlier. The first emergence of structure is usually related to what is called the thermodynamic branch. When some parameter is changed which represents the move away from thermodynamic equilibrium, that equilibrium solution can be 'continued' and for a time it remains the most likely state. The energy is still distributed randomly and there is no structure. However, at some critical point, the thermodynamic branch may become unstable and new stable steady-state solutions emerge. Fluctuations are then amplified and take the system to the new steady state. This is illustrated in an idealised form in Figure 8.2. These new states may then exhibit macroscopic structures. These may be temporal, as with the cycles of the previous subsection, or spatial, or both. Hence: 'order from fluctuations'. We illustrate these ideas by reference to the famous Belousov-Zhabotinsky reaction.

One of the main forms of this reaction involves malonic acid and bromates and Field and Noyes (1974), Field (1975) have shown that a suitable simplified model of its takes the form

$$A + Y \xrightarrow{k_1} X \tag{8.13}$$

$$X + Y \xrightarrow{k_2} P \tag{8.14}$$

$$B + X \xrightarrow{k_{34}} 2X + Z \tag{8.15}$$

$$2X \xrightarrow{k_5} Q \tag{8.16}$$

$$X \xrightarrow{k_6} fY \tag{8.17}$$

The openness of the system is achieved by fixing A and B and it is then possible to focus on the concentrations of X, Y and Z. The differential equations which follow from (8.13)-(8.17) are

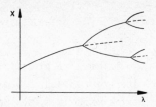

Figure 8.2 Chemical concentration vs. parameter bifurcations

$$\dot{X} = k_1AY - k_2XY + k_{34}BX - 2k_5X^2 \qquad (8.18)$$

$$\dot{Y} = -k_1AY - k_2XY + fk_6Z \qquad (8.19)$$

$$\dot{Z} = k_{34}BX - k_6Z \qquad (8.20)$$

One interesting feature is that the system has to be analysed
on at least two time scales because of the widely differing
magnitudes of the constants in the equations.

It can be shown that there is an area with a single steady-
state solution determined by a parameter w given by

$$w = k_6/(k_1k_{34}AB)^{\frac{1}{2}} \qquad (8.21)$$

In the region

$$0 < w < w_c \qquad (8.22)$$

for some w_c, the single solution becomes unstable and there is
an alternative periodic solution.

In the examples considered so far, the equations have not
contained a spatial diffusion term.   This is often necessary
because the concentrations of A and B can only be maintained by
interactions at the boundary.   When this is done, it turns out
that spatial macroscopic structures are found.   In the case of
the Belousov-Zhabotinsky reaction, these take the form of spiral
waves around a centre.

Chapter 8

8.3 Biology

8.3.1 Introduction

There has been a wide range of applications of catastrophe theory and bifurcation theory in biology. Some of these are concerned with very specific mechanisms, such as Zeeman's work on the nerve impulse and the heartbeat and since they do not appear to offer any insights for geographical problems we do not consider them further here. The questions and tasks which may be of interest concern mainly the development and evolution of structure and pattern in biological systems. These questions have been considered at a variety of scales, from the sub-cellular to the whole organism. We sketch briefly below the different kinds of ideas which have been used, firstly to model the development of structure and secondly evolution. For more extensive surveys and references, the reader is referred to the appropriate sections of the books by Zeeman (1977-A), Poston and Stewart (1978) and Nicolis and Prigogine (1977).

8.3.2 Developmental biology

Organisms are made up of interacting cells, and cells can be considered as mixtures of interacting chemicals. The development of an organism is controlled in some way by its chromosomes or sets of genes. These are meso-molecules which contain the information necessary for the development of the organism as a whole. The effect of genes is catalytic: they can generate the production of certain enzymes which are then used in protein building.

At the sub-cellular or cellular scales, because the main interactions are chemical, the mechanisms sketched in a different context in Section 8.2 are appropriate. For example, at the sub-cellular scale, the reaction in the neighbourhood of a particular gene can be modelled. The full range of bifurcation properties of the corresponding dynamical equations are useful in different ways: there may be the cyclic production of particular molecules, for example; in another case, the model

shows steady production of an enzyme until a certain quantity is
generated after which production is inhibited.

At the cellular level, the whole set of reactions involved
is obviously much more complicated: the whole set of genes is
involved. In this case, simpler methods of analysis have been
used, in particular the notion of 'genetic switching'. A gene
is involved in a chemical reaction which produces a particular
protein, but it does not function in this way all the time. It
can be either 'on' or 'off'. Whether it functions or not
depends on its own chemical environment at a particular time.
In other words, it depends on the presence or absence of other
proteins which can be considered as inputs to that gene - these
proteins being the outputs of other genes and its own protein
product being its own output which in turn is an input for other
genes. The set of genes in a cell can thus be regarded as
forming a connected net (see, for example, Kauffman, 1969, Glass
and Kauffman, 1973). The simplification of method which is then
introduced into the analysis is to assume that the off-on state
of each gene can be characterised by a 0-1 Boolean variable.
Kauffman then makes a further theoretical step: to assume that
the connections in the net are *random*. In turns out that, even
with this assumption, stable or cyclic states can be identified.
This means that explanations of biochemical behaviour in cells
can be derived from rather weak assumptions: that the programme
which brings about certain states depends on the Boolean logic
rather than on some specific spatial arrangement. This is the
only method so far which has been applied directly in urban
modelling and we give a more detailed account of the principles
of it in Section 8.6 below. The 0-1 characterisation of the
on-off switch is clearly likely to be an approximation, and
Poston and Stewart (1978) argue that a cusp-catastrophe model of
such switching may be appropriate. This would add the possible
additional complication of hysteresis effects.

We saw in our brief discussion of the Belousov-Zhabotinsky
reaction above that spatial patterns can occur in interacting
chemical mixtures. This is obviously an important feature of

233

cell and multi-cell dynamics in the study of morphogenesis in
embryology. The possibility of such structures - for groups of
cells - arising from chemical reactions was proposed by Turing
(1952) and a variety of models have been proposed since. It is
difficult to make much progress mathematically and realistically
because diffusion terms need to be added to the differential
equations which already are complicated because of the number of
variables involved. Further, the solution of such equations
would be immensely difficult because of the complicated boundary
conditions determined by the shape of the cell membranes.

Another kind of simplification is to look at the interaction
of particular kinds of chemicals in the organism as a whole.
Sprinkhuizen-Kuyper (1976), for example, has looked at antigen-
antibody reactions and, perhaps not surprisingly, develops in
yet another context forms of the Lotka-Volterra prey-predator
equations.

Finally, we note a direct application of catastrophe theory
to morphogenesis by Cooke and Zeeman (1976). This involves the
notion of a 'travelling wave' as the basic mechanism. However,
since any application of this idea in geography is likely to
have closer analogies with ecology than embryology, and it also
has been applied in that field, we reserve a more detailed
discussion of it to Section 8.4 below.

8.3.3 Evolutionary biology

The modern theory of evolution is a mixture of the
principles of Darwinism and the results from molecular biology
discovered in the last twenty years. Since the programme for
the development of a particular organism is contained in its
genetic structure, the problem of understanding the evolution
of a new species becomes one of understanding the ways in which
this genetic structure can change (see Mayr, 1978 and Ayala,
1978, for recent reviews). Evolution is considered as a two-
step process. The first is the generation of genetic variation
(by what Mayr describes as 'recombination, mutation or chance
events', and the second is some kind of selection process.

The details, at this stage at least of the development of
geographical theory, are probably not of sufficient consequence
to develop in detail.

The nature of the problem, however, is directly interesting
to us, and there are two features of particular importance. The
first is: how do systems of considerably higher complexity and
organisation than previously existing systems evolve? The
second is: how does the particular path of the evolution of
organisms and systems of organisms evolve in interaction with
their environments (May, 1978, Maynard Smith, 1978)?

The first question is probably related to some of the basic
issues of general systems theory and will involve concepts of
subsystems and suprasystems at various levels in a hierarchy.
The second can be partly tackled through the concept of an
ecological niche, though here there is the difficulty that the
definition of a niche for a particular organism depends on the
pattern of existence of other (competing) organisms. (This seems
rather analogous to, if more difficult than, the W_k, $k \neq j$, backcloth problem when analysing one particular centre, W_j.) There
have been attempts to model genetic evolution, as a result of
competition, with dynamical systems models. These are elaborations of the competing species' models of Chapter 2 with the
genetic components of the population made explicit - see, for
example, Levins (1968), Levins (1971), Leon (1974) and Allen
(1976). It has also been argued that dissipative structures
play a role (Prigogine, Nicolis and Babloyantz, 1972).

It is an immensely difficult task to answer the question:
how did the species of a particular system evolve. It is a
very much harder task to predict the path of evolution in the
future. For urban systems, the pace of evolution is much more
rapid than for biological systems. New kinds of structure and
organisation have evolved very rapidly, and especially in recent
years with technological change. We discuss the nature of this
kind of evolutionary question further in the geographical
context in Section 8.6 below.

Chapter 8

8.4 Ecology

8.4.1 Population dynamics: differential equations

One of the main fields of ecology is the study of the growth and decline of populations of different species, particularly those interacting with each other. An elementary account of the differential equations which can be used was offered in Chapter 2 to illustrate basic concepts and bifurcation. The reader who is not familiar with those examples should perhaps recapitulate on Chapter 2 before proceeding. We also introduced very briefly the notion that difference equations are sometimes more appropriate than differential equations. We use this section, in part, to explore more deeply the essential distinctions between these types of equations which in part arise out of practice - the numerical integration of differential equations - and in part out of a different underlying model. First, however, we review developments with differential equation models.

The most widely used equations are versions of the Lotka-Volterra equations, either in their prey-predator form or in the competition-for-resources form. They have been applied in a very wide variety of situations. In the previous subsections we noted applications in chemistry and biochemistry. Here, we concentrate mainly on interactions between animal species, but occasionally refer to examples where the prey is vegetation and the predator is animal. We begin by examining a general formulation of the Lotka-Volterra equations and then outline a number of developments, which involve adding different aspects of real-world complexity to the basic models.

A convenient general form of the equations is

$$\dot{N}_i = r_i N_i \left(\frac{K_i - \alpha_{ii} N_i - \sum\limits_{j \neq i}^{n} \alpha_{ij} N_j}{K_i} \right) \qquad (8.23)$$

for n species, populations N_i and constants r_i, K_i and α_{ij} (which have the roles of growth rates, carrying capacities and

competition or inhibition coefficients respectively). The
equilibrium solution is

$$N_i^* = (K_i - \sum_{j \neq i} \alpha_{ij} N_j^*)/\alpha_{ii} \qquad (8.24)$$

The stability properties of these equations can be studied by
standard methods of linear algebra. In typical cases, they
generate a set of relations between the coefficients which have
to be satisfied for stability. The increasing complexity of
these with n leads May (1973) to the hypothesis that as systems
become increasingly complex, they become less stable: there is
no obvious *a priori* reason why so many conditions should
regularly be satisfied. A number of special cases have been
studied. A good example is the antisymmetric case:

$$\alpha_{ij} = -\alpha_{ji} \qquad (8.25)$$

which implies

$$\alpha_{ii} = 0 \qquad (8.26)$$

Since α_{ii} cannot be zero in (8.24), the equilibrium populations
are now found as the solutions of

$$\sum_{j \neq i} \alpha_{ij} N_j^* = K_i \qquad (8.27)$$

Again, there are conditions on the coefficients for the existence
of stable equilibria and it can be shown that the system exhibits
purely oscillatory behaviour when it is disturbed. There is one
conditions which in many eyes (*cf*. May, 1973, p. 53) effectively
discredits the antisymmetric model: stable equilibria only
exist when there is an even number of species.

The prey-predator model and its multispecies extensions
essentially represent interactions between two trophic levels.
May (1973), following Gardner and Ashby (1970), investigates the
stability of systems with randomly connected food nets - in a
manner rather analogous to the randomly connected genetic nets
mentioned earlier. (There is also an application of the Lotka-
Volterra equations to neural nets in Cowan (1970).) This

produces some interesting results, derived from exploring stability: for example, that if a species interacts with many others it should do so weakly and vice versa. It is argued by Levin (1970), for example (quoted by May) that the "dynamics of a broad class of complex systems will result in simplification through instability". In much of this work in effect, the prey-predator idea and the competition-for-resources concept are combined.

One difficulty in ecological work, which also occurs in many other disciplines, is that detailed knowledge of the coefficients in the equations is often unavailable. However, the signs of interactions are usually known and this means that digraph methods can be used to investigate some stability properties. That way, at least some systems can be identified as unstable without the detailed knowledge. A detailed treatment of such methods is offered by Roberts (1976).

We have seen in earlier contexts that, particularly with a view to geographical applications, it is important to add spatial diffusion terms to the models. A good example of this is provided by Steele (1974) who models the interactions between phytoplankton (labelled P, the prey) and zooplankton (labelled H, the predator). The Lotka-Volterra equations with diffusion added then become:

$$\frac{\partial P}{\partial t} = aP - b(P - P^*)H + \frac{\partial}{\partial x} k \frac{\partial P}{\partial x} \tag{8.28}$$

$$\frac{\partial H}{\partial t} = c(P - P^*)H - dH + \frac{\partial}{\partial x} k \frac{\partial H}{\partial x} \tag{8.29}$$

where a, b, c, d and k are constants. P^* is another constant which is a 'grazing threshold': grazing does not take place until P exceeds P^*. His solutions then include spatial as well as temporal oscillations.

An alternative and simpler treatment of some spatial effects is offered by Pielou (1974). She uses the kinds of results we presented in Chapter 2 on stability in a two species competition-for-resources model to show how the mix of species can change because of the existence of an environmental gradient. If this gradient is reflected in one or more of the coefficients, then this could lead to a situation where there is an area in which one species is dominant, followed by a co-existence area, followed by an area in which the other species is dominant.

Two further kinds of developments will be mentioned briefly in the context of differential equations, though both can be taken further using difference equations. First, some processes involve time lags. May (1973) gives an example which is yet another modification of Lotka-Volterra equations for herbivore (H) and carnivore (P) populations:

$$\dot{H}(t) = rH(t)[1 - H(t - T)/K] - \alpha H(t)P(t) \qquad (8.30)$$

$$\dot{P}(t) = -bP(t) + \beta P(t)H(t) \qquad (8.31)$$

In this case, there is a time delay on the effect of resource limitation in the growth of the prey population. A characteristic time can be defined for the system as

$$\tau = (rb)^{-\frac{1}{2}} \qquad (8.32)$$

which is the geometric mean of the times associated with birth-rates and death-rates of the prey and predator populations respectively. Then the equilibrium points of Equations (8.30) and (8.31) is stable or unstable according to whether

$$\begin{array}{ll} \tau > T & \text{stable)} \\ & \qquad) \\ \tau < T & \text{unstable)} \end{array} \qquad (8.33)$$

The need for age disaggregation arises when there are periodic fluctuations in the environment of a period considerably less than the life-time of a species in the system. A variety of models to meet these situations are presented by Oster and Takahashi (1974). They can be developed in a difference equation format and are probably less important than some of the other developments for geography since age-disaggregation in geographical models is more common anyway (*cf*. Rees and Wilson, 1977).

8.4.2 Population dynamics: difference equations

May (1974, 1975, 1976-A, May and Oster, 1976) has presented a number of striking results on complex dynamic behaviour associated with very simple difference equations. This provides us with a new perspective which turns out to have direct applications to some situations in urban geography. First, however, we comment generally on the distinctions in use between differential and difference equations both in ecology and more widely. If the main state variables are changing continuously, then differential equations are appropriate; if the events can be considered to be discrete, then difference equations offer the correct formulation. In ecology, if elements of populations typically have lives longer than the period of analysis and associated events, then differential equations are appropriate. But often, a species will all die out by the end of a single time period, leaving, say, eggs to create the next generation in the next time interval. These are known as non-overlapping generations. Difference equations are then most appropriate. It should also be added that when differential equations are integrated numerically, they are in effect being treated as difference equations and some of the results described below are applicable.

May's argument can be illustrated with reference to very simple difference equations of the form

$$X_{t+1} = F(X_t) \tag{8.34}$$

for a single population, X and a suitable function F. In his review paper (May, 1976-A), he concentrates on the following particular example:

$$N_{t+1} = N_t(a - bN_t) \tag{8.35}$$

whose form suggests that it is a discrete version of the logistic growth (differential) equation presented in Chapter 2. It can be written in the form

$$N_t = N_{t+1} - N_t = b\left[\left(\frac{a-1}{b}\right) - N_t\right]N_t \qquad (8.36)$$

to make the resemblance more explicit. b is then the rate constant and $(a - 1)/b$ the carrying capacity.

If we make the transformation

$$X = bN/a \qquad (8.37)$$

then Equation (8.35) can be written in the canonical form

$$X_{t+1} = aX_t(1 - X_t) \qquad (8.38)$$

X can then be considered to vary between 0 and 1. The only problem then is that if X ever exceeds 1, the sequence diverges to $-\infty$ and the negative numbers are unrealistic in most applications. This can always be avoided by the use of an alternative form of (difference) logistic equation:

$$X_{t+1} = X_t \exp[r(1 - X_t)] \qquad (8.39)$$

However, we ignore the complication of X negative here and assume the constants and initial conditions are such that it does not arise. The nature of the results is the same in both cases. So we work with Equation (8.38).

The relation between X_{t+1} and X_t can be plotted as a humped curve as shown in Figure 8.3. It attains a maximum at $X = \frac{1}{2}$ of $a/4$, and since X must remain less than 1, this imples $a < 4$. We also require $a > 1$ to avoid $X_{t+1} \rightarrow 0$ for large t. So, for non-trivial dynamic behaviour:

$$1 < a < 4 \qquad (8.40)$$

The possible equilibrium values of X are found by putting $X_{t+1} = X_t$ in Equation (8.38) and this is equivalent to seeking the intersection of the humped curve in Figure 8.3 with the 45^0 $X_{t+1} = X_t$ line. This is also plotted on Figure 8.3 and thus

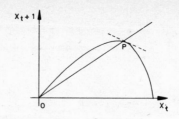

Figure 8.3 Logistic difference equation X_{t+1} vs. X_t

the point P and the origin are equilibrium points. We now explore the stability of these points.

Let $X = X^*$ be the non-zero equilibrium point. For later notational convenience, we also write Equation (8.38) in the form (8.34) with

$$F(X) = aX(1 - X) \tag{8.41}$$

At equilibrium

$$X_{t+1} = X_t = X^* \tag{8.42}$$

and so Equation (8.38) then gives

$$X^* = aX^*(1 - X^*) \tag{8.43}$$

which has (non-zero) solution

$$X^* = (a - 1)/a \tag{8.44}$$

The slope of the curve at this point is

$$\frac{\partial F}{\partial X}\bigg|_{X = X^*} = a - 2aX^* \tag{8.45}$$

which is

$$\frac{dF}{dX}\bigg|_{X = X^*} = 2 - a \tag{8.46}$$

(Substituting from (8.44).)

Consider points $X^* \pm \Delta$ within a small increment, Δ, of the equilibrium point, as in Figure 8.4. If the slope of the tangent is between ± 1, and if we consider the topside point

$$X_t = X^* + \Delta \tag{8.47}$$

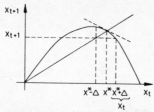

Figure 8.4 Logistic difference equation: stability of equilibrium solution

then the figure shows that

$$X^* \; < \; X_{t+1} \; < \; X_t \tag{8.48}$$

which shows that the equilibrium is stable. Thus P is a stable equilibrium point if the gradient, 2 - a, given in Equation (8.46) is between ±1. This implies

$$1 \; < \; a \; < \; 3 \tag{8.49}$$

for stability. The same kind of analysis applies at the origin. Putting $X^* = 0$ in Equation (8.45) shows that the gradient at the origin is a. This is an unstable point if a > 1. It is only stable when a < 1 which is the trivial case when $X_t = 0$ for all t. In this case, the upper point of intersection, P, does not exist. This is just a restatement of the earlier result that we need a > 1 for non-trivial behaviour.

What happens when a exceeds 3 and P becomes unstable?   In Figure 8.3, as a increases, the hump steepens and it is easy to see that a point will be reached when the modulus of the tangent at P exceeds 1.   This occurs when a = 3. The next step is to explore the possibility of an equilibrium point two time periods past;   that is

$$X_{t+2} \; = \; F[F(X_t)] \tag{8.50}$$

If X_{t+2} is plotted against X_t, the curve has two humps. Three cases are shown in Figure 8.5. Case (a) has a < 3. This shows the stable one-period equilibrium point (which also obviously exists and is stable over two intervals).   Case (b) shows the 45[0] line touching the curve;   this represents the limiting case

243

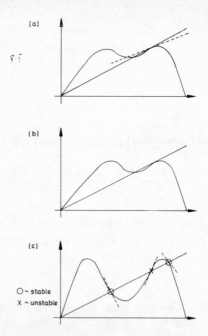

(a)

8.5

(b)

(c)

O ~ stable
X ~ unstable

**Figure 8.5 Logistic difference equation: stability of
one-period oscillation**

$a = 3$.   The $a > 3$ case is shown in case (c).   The tangent at
the original stable equilibrium point now exceeds 1 and is
unstable, but there are now two new stable equilibrium points,
$x^{(2)}*$ and $x^{(2)}**$. $a = 3$ is therefore a bifurcation point, say
$a_c^{1/2}$.   This situation holds up to another critical value which
we can label $a_c^{2/4}$ when, as the notation implies, only four
period stable points exist.   For $3 < a < a_c^{1/2}$, the system
oscillates between four stable points.   This '2^n' sequence continues up to $a = 3.57$.   For $3.57 < a < 3.8495$, more cyclic
behaviour is identified, made up of sets of 3-cycles, but for
$a > 3.8495 = a_{ch}$, say, the behaviour is oscillatory but chaotic:
there is no discernable period.   As we noted earlier, the limit
of this behaviour occurs at $a = 4$.   This is all represented
diagrammatically in Figure 8.6.

The most striking feature of these results is that such
complex behaviour can arise from such a simple equation.   There
244

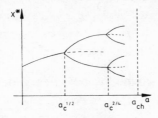

Figure 8.6 Logistic difference equation : bifurcation pionts

are direct applications to our shopping centre/urban structure
example which will be pursued in Section 8.6 below. It only
remains here to note that the argument can be extended to more
complicated situations and that then we can expect even more
complicated bifurcation behaviour to emerge. Beddington, Free
and Lawton (1975), for example, show how such complex behaviour
arises in Lotka-Volterra prey-predator equations when they are
cast as difference equations. An explicit formulation of such
a difference model is offered by Hassell (1978) as

$$N_{t+1} - N_t = N_t(\alpha - bP_t) \tag{8.51}$$

$$P_{t+1} - P_t = P_t(-\beta + cN_t) \tag{8.52}$$

(using an obvious notation). This turns out to be essentially
the same as the Nicholson-Bailey host-parasite models which
were formulated in difference equation terms. Again, relations
between the parameters can be obtained as stability conditions.

A similar presentation is offered by Innis (1974) using,
as a basis, Forrester's (1968) systems dynamics methods. He
draws particular attention to the different results obtained
with the difference and differential equation formulations and
how the closed orbit of the latter can become a spiral (in
either direction) in the former. He discusses this in terms of
the step-length needed for effective numerical integration of
differential equations.

The final complication to be noted briefly refers to the
lag structure implicit in difference equations. In the first

245

order models presented above, the situation at time t is respon-
ded to at time t+1.   If some effects are not registered until
t+2, this leads to second order equations, and so on.

8.4.3 Travelling waves in ecological models

For our final example in ecology, we turn to a very different
style of work based on a more classical catastrophe theory
approach.   Consider a situation similar to the one mentioned in
relation to some of Pielou's work above: the existence of an
environmental gradient such that one end is suitable only for
development as grassland, the other only as forest.   Zeeman (1974)
shows that, under certain conditions, a sharp frontier is established between the two types of vegetation, but this frontier
is first established elsewhere and travels to its final stopping
place, deepening as it does so, as a wave.

If the environmental gradient is characterised by a single
spatial dimension, then the result which is to be proved is
characterised by Figure 8.7.   In effect, what Zeeman does is
to raise a third (perpendicular) axis on this figure and to
generate an equilibrium state surface in this three-dimensional
space.   This surface represents possible equilibrium vegetation
states and he shows that, typically, it will have two sheets,
the upper one representing, say, grassland, the lower one forest.

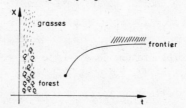

Figure 8.7 Enviromental gradient and travelling frontier

The assumptions used are that the system has the following
properties: homeostasis, continuity, differentiation and
repeatability.   Homeostasis is interpreted as meaning that the
system state tends to one of (in this case) two stable states,
grassland or forest and that this is achieved by the minimisation

of a potential function. Continuity refers to the parameter
change in the environmental gradient and differentiation to the
existence of the two stable states. Repeatability (of experiments which 'satisfy' the model) guarantees that the surface of
equilibrium states is generic - that is, does not have any
untypical properties.

The above assumptions together imply from Thom's classification theorem that the surface of possible equilibrium states
will have singularities which are at worst fold curves and
cusp points. Differentiation implies that the folds exist in
this case, and repeatability that there will be a cusp point.
The shifting frontier point in Figure 8.6 is then the projection
of the fold part of the surface of equilibrium states, and this
is sketched in Figure 8.8.

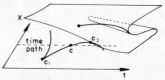

Figure 8.8 Cusp catastrophe explanation of travelling function

This model represents an example of the catastrophe theory
model offering a description of a mechanism without any detail.
Poston and Stewart (1978) write about this idea (in this
particular context in fact) as follows:

> ... extend the rule of thumb of all the sciences 'Try
> a linear model first', when bifurcation phenomena make
> clear that this won't work, to 'Try a catastrophe
> model next'.

So this principle is illustrated here, and also the idea of what
happens when space and time coordinates are used as control
variables in a catastrophe model. Thom (1975) always saw this
as one of the main areas of application of catastrophe theory,
particularly in embryology and Zeeman (1974) and Cooke and
Zeeman (1976) have worked out detailed models in this field.

We noted above that the problem was similar to that posed
by Pielou (1972), but the result is different and this perhaps

needs comment.   In her case, at least when the coefficients
were such that a stable co-existence equilibrium was possible,
there was an area (for Zeeman's example) of forest, followed by
a mixture, followed by an area of grassland.   The frontier was
not sharp.   The model, however, was based on Lotka-Volterra
'competition' equations and these, while derivable from a
Hamiltonian (Kerner, 1972) do not arise from potential functions.
Nor does the notion of *sharp* differentiation always hold.   So
the model is simply a different one.   It remains, as always, a
matter of empirical investigation to see which model applies in
which cases.

8.5 Economics

8.5.1 Introduction

Economists seem, to date and relatively, to have neglected
the possible contributions of catastrophe theory and bifurcation
theory at least as reflected in mathematical literature.   This
seems surprising as so many economic models could perhaps be
illuminated from this new perspective.   This is because of a
long-standing concern with such concepts as equilibrium,
stability and dynamics in economic models.   Some results are
already built into the literature, of course.   For example,
there is an extensive treatment of different kinds of solutions
to difference equations and their stability (cobweb theorems,
eg. see Baumol, 1959).

There has also been a recognition of the existence of
multiple equilibria in some models (for example, Samuelson,
1947, pp. 49 and 240).   Smith (1977) has explored a model of
the location of banks which generates multiple solutions and
discrete changes.   Explanations have recently begun on the
relationships between such multiplicity and catastrophe theory
(Balasko, 1975, 1978).   There are also connections to game
theoretic models, perhaps well illustrated from outside economics
by Richardson's (1960) well-known work on the mathematics of
war, where competing nations behave according to equations which

are like modifications of the Lotka-Volterra equations (Richardson, 1960, p. 56).

In this section, we consider two examples, the first concerned with the management of renewable resources, where the economics is linked to ecological models, and the second concerned with business cycles.

8.5.2 Resource management

In this subsection, we mainly follow and summarise some of the arguments of Clarke (1976) concerned with fisheries management. It is interesting because it is a different kind of example to any considered hitherto and also offers a different kind of technique. This has similarities with the difference equations methods of May reported above (and indeed those methods have a direct application here), but are first formulated in continuous variable terms and focus on plots of $\dot{x}$ against x for the main 'stock' variable, x. This therefore offers methods based on the style of plot shown in Figure 2.6 of Chapter 2.

The main variables and concepts can be introduced briefly as follows. Let x be the resource population at time t and h(t) the rate of removal. Let F(x) be a function representing the natural growth of the population. Then

$$\dot{x} = F(x) - h(t) \tag{8.53}$$

Let E be the amount of effort devoted to fishing in units so that the removal rate can be taken as

$$h = Ex \tag{8.54}$$

that is, proportional to both the effort and the stock. Equation (8.53) then becomes

$$\dot{x} = F(x) - Ex \tag{8.55}$$

Let x* be an equilibrium point of this equation, and then the substainable yield, Y, is given by

$$Y = Ex^* \tag{8.56}$$

Chapter 8

Suppose now that $F(x)$ is represented by a logistic function:

$$F(x) = rx(1 - x/K) \qquad (8.57)$$

for suitable constants r and K. Then the equilibrium occurs when

$$rx^*(1 - x^*/K) = Ex^* \qquad (8.58)$$

which gives

$$x^* = K(1 - E/r) \qquad (8.59)$$

with yield

$$Y = KE(1 - E/r) \qquad (8.60)$$

(using (8.56)). These results are shown graphically in Figure 8.9. The dashed curve is the plot of $\dot{x}$ against x, and its components, $F(x)$ and Ex are shown separately. The yield-effort (Y - E) curve, from Equation (8.60), is plotted in Figure 8.10. This shows that the maximum sustainable yield occurs when $E = r/2$. If effort is increased beyond this level, then the yield decreases smoothly until it reaches zero at $E = r$.

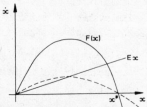

Figure 8.9 Natural growth and harvesting functions for a fish population, and their difference

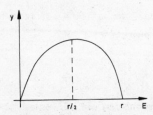

Figure 8.10 Yield as a function of effort

F(x) in Figure 8.9 is an example of what is called a pure
compensation curve, where the rate per unit, $F(x)/x$, is always
a decreasing function of x. An important class of models
arises when F(x) is a depensation curve, when $F(x)/x$ is an
increasing function of x for some range $0 < x < K^*$. It exhibits
critical depensation when $F(x) < 0$ for some small values of x,
say $0 < x < K_0 < K^*$. K_0 is called the minimum viable population level. Examples of such curves are shown in Figure 8.11.
The line through the origin which is a tangent to the curves
has gradient E^*.

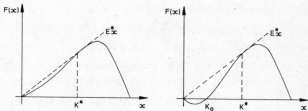

Figure 8.11 Alternative natural growth functions

The basic diagrams for the non-critical depensation case
are presented as Figure 8.12 - (a) is the x-x plot, (b) is the
yield-effort plot. The system now exhibits a number of interesting features which were not in the pure compensation model.
There is one common feature: as E increases from a low level,
there is an equilibrium point x* and a corresponding yield Ex*.
This reaches a maximum, at E^M say. Suppose now E is increased
further. For $E > E^*$, the stable equilibrium point x* no longer
exists and the yield jumps to zero. This is shown in Figure 8.12
(b) and is a clear example of the fold catastrophe.

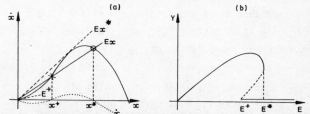

**Figure 8.12 Differences of natural growth and harvesting functions
for alternative growth functions**

Let us now investigate the other equilibrium points shown on Figure 8.12(a). x^+ is always unstable and generates the (unstable) yield which is the dashed part of the yield-effort curve. The origin is stable for $E > E^+$, where $E^+ = F'(0)$, the gradient of $F(x)$ at the origin. (And x^+ only exists when $E > E^+$). So consider what happens if fishing effort is increased continuously. When E^M is passed, this may not be immediately noticely noticed and if E passes E^*, then yield will suddenly drop to zero. If E is decreased again, then if $E^M > E^+$, the origin is a stable equilibrium point and the situation cannot be recovered by this reduction. If E is reduced to a level less than E^+, however, then the origin becomes an unstable equilibrium point and x can increase, and then E can slowly be increased to E^M again. Thus, there is a hysteresis effect which is shown explicitly on the yield-effort curve in Figure 8.13. This model has some similarities with the shopping centre size model of Chapter 5 in that the yield-effort curve is a fold curve but with zero states added. The multiple solutions in the region $E^+ < E < E^*$ generate the hysteresis.

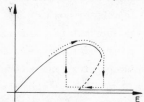

Figure 8.13 Yield vs. effort and a fold catastrophe

In the critical depensation case, the same effects are shown except that since the origin is a stable state for all E (>0, <0 would imply restocking), once yield goes to zero for $E > E^*$, it stays there. The change is irreversible. This can be seen in an argument similar to that contained in Figure 8.12 and this is left as an exercise for the reader.

There are obviously many ways in which this model can be elaborated - both by incorporating alternative ecological models and by extending the economic framework. In the latter case, for example, the costs of different kinds of effort can be incorporated to allow for scale economies or whatever. The reader is referred to Clark's book for such elaborations and to the article by Hoppensteadt (1978) for an account of difference equation formulations which relate closely to our presentations of May's work earlier.

8.5.3 Business cycles

May (1976-A) noted the potential relevance of his results, on complex dynamic behaviour arising from simple difference equations, to the theory of business cycles and he quoted Goodwin (1951), who was one of the first economists to seek to introduce non-linearities into dynamic economic models in order to generate such behaviour. Here, however, we concentrate on the approach of Varian (1979) who both introduces non-linearities and links these to the cusp catastrophe. This allows him to have either fast or slow returns from a depression or recession as we will see. His dynamic model is

$$\dot{y} = S[C(y) + I(y,k) - y] \qquad (8.61)$$

$$\dot{k} = I(y,k) - I_0 \qquad (8.62)$$

y is national income; C, consumption; I, gross investment; and I_0, 'replacement' investment. S is a parameter measuring the speed of response of income adjustments (which are supposed to be much faster than those concerned with investment). The crucial decisions are the forms of the functions $C(y)$ and $I(y,k)$. A linear assumption is made for $C(y)$:

$$C(y) = Cy + D \qquad (8.63)$$

A savings function is defined as

$$S(y) = y - C(y) = (1 - C)y - D \qquad (8.64)$$

$I(y,k)$ is assumed to grow logistically with y, and to decrease with increasing k. The equilibrium points for Equation (8.61) occur when

$$S(y) = I(y,k) \qquad (8.65)$$

253

and examples of the various curves and the equilibrium points are plotted on Figure 8.14. This shows that there can be one low stable equilibrium value of y, one high stable equilibrium value, or three equilibrium values, the outer two stable, the inner one unstable. This generates the plot of $\dot{y}$ = 0 on Figure 8.15 and it contains folds. The plot of $\dot{k}$ = 0 can be shown to be an upward sloping straight line and Varian proves that this intersects the $\dot{j}$ = 0 curve once and once only. He also shows that an intersection on the middle part of the curve generates cyclic behaviour.

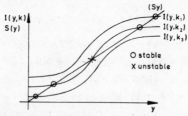

Figure 8.14 Investment and savings functions vs. national income

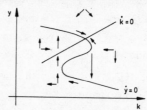

Figure 8.15 Plots of $\dot{y}$ = 0, $\dot{k}$ = 0, in state space

The explanation of recessions can then be cast in terms of Figure 8.15. A shock can send the system from the upper curve to the lower curve. There is an adjustment back along this curve and then a jump back to the upper curve. He defines a depression on the other hand as a fall from which there is a large and slow recovery. To widen the model, he introduces wealth, w, as an additional control variable and makes C, and hence S, a function of w. Equation (8.63) becomes

$$C(y,w) = C(w)y + D(w) \qquad (8.66)$$

and

$$S(y,w) = y - C(y,w) \qquad (8.67)$$

This can change the slope of the S(y,w) line (for low w as we will see shortly) in Figure 8.14 to that shown in Figure 8.16. For 'intermediate' values of k and w, the behaviour is as before and folds are generated.

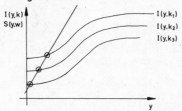

Figure 8.16 Alternative savings function, with investment functions

If wealth becomes small, however, S(y,w) shifts upwards and becomes more steep.   Then no shifting of k will generate three equilibria as shown in Figure 8.16.   This means, in effect, that while k acts as a normal factor, w acts as a splitting factor.   The resulting possible equilibrium states take the familiar cusp form as shown in Figure 8.17.   A depression is then a recovery from a decline which takes place smoothly with a setting of the w-factor which keeps the trajectory clear of jumps.

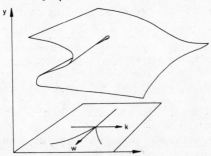

Figure 8.17 Depression and the cusp catastrophe

Varian's model then offers various solutions: stable y, high or low; business cycles generated by the 'middle curve'; and with an extended model, the possibility of slow returns as well as fast ones.

8.6 Applications to cities

8.6.1 Introduction and review

The fields reviewed above are mostly concerned with interacting populations, or populations with interacting elements. We have seen that such systems can be modelled by non-linear differential or difference equations and that often the same or (topologically) similar equations are applicable in a wide variety of circumstances - the ubiquitous Lotka-Volterra equations being a good example. We have also seen that a considerable amount of insight can usually be gained by geometrical arguments and some of the examples in this chapter have added to our repertoire in this respect.

What can we learn from the experience of other disciplines about the geography of cities? The answer to this question comes in two parts. First, there is enough similarity of type of problem to give us some confidence that theoretical geography is now functioning at the frontiers of science in general. Indeed, many of the non-linearities and associated complications (eg. the $\{W_k\}$, $k \neq j$, backcloth) of geographical models suggest that some of the geographical problems are harder than those in some other disciplines. Secondly, by investigating methods in other disciplines, it is still possible to obtain new ideas for geographical modelling. We offer two examples of this below, the first the use by some members of the Brussels school of the genetic switching idea; and the second, the application of May's ideas on difference equations to the dynamics of shopping centres.

8.6.2 Urban models and Boolean algebra: analogues of genetic switching

The ideas on genetics on which these urban models are based are reported, for example by Kauffman (1969) (and see also Glass and Kauffman, 1973, Kauffman, 1977, Thomas, 1978). They have been applied in the urban context by Boon and de Palma (1978) and de Palma, Stengers and Pahaut (1979), whose arguments we follow here.

The model is best explained using the example of Boon and de Palma. They develop a model of residential location with four sets of groups of residents, labelled A, B, C, D. Two kinds of variables are defined: a, b, c, and d which are decision variables and α, β, γ, δ which are 'memory' variables. These can each be zero or one and are combined by the standard rules of Boolean algebra. A value of 1 indicates the presence of that group in an area, 0 its absence. In effect, the α, β, γ and δ variables are the values of a, b, c and d for the previous time period. The model structure takes the form, for the a and α variables, for example:

$$\alpha_t = a_t \tag{8.68}$$

$$a_{t+1} = a(\alpha_t, \beta_t, \gamma_t, \delta_t) \tag{8.69}$$

for time subscripts t, and with similar equations for b, c and d in relation to β, γ and δ. We see, therefore, that the model equations are first order difference equations using 0-1 variables and Boolean logic. It is also easy, as with difference equations in general, to incorporate a more general lag structure.

In the model used as an example here, A is the highest income group through to D, the lowest. Four additional variables have to be defined before we can write down the model equations: T, H and D are measures of travel to work costs, environmental quality and density respectively, assumed for now to be given exogenously for different zones, and p is a house price variable (with 1 meaning 'high'; 0, 'low'; and Π the associated memory variable).

Assume that high income people want a good environment (H), not to mix with lowest income people ($\overline{\delta}$), they will accept the third group provided some of their own class are already there ($\alpha + \overline{\gamma}$), and they will not accept both high density and second-group people together ($\overline{D.\beta}$). This leads to the equation:

$$a = \overline{\delta}.H.(\overline{D.\beta}).(\alpha + \overline{\gamma}) \tag{8.70}$$

257

Similar arguments are given for the other groups:

$$b = \overline{\delta}.(\overline{D.\gamma}).(\overline{\Pi.T} + \alpha) \tag{8.71}$$

$$c = \overline{\delta}.\overline{\Pi}.T \tag{8.72}$$

$$d = \delta.\overline{\Pi}.T \tag{8.73}$$

and the price equation is

$$p = \overline{\gamma}.\overline{\delta}.(\alpha + H.\overline{D}) \tag{8.74}$$

It is argued that the exogenous variables, T, H and D are either truly exogenous or can be fixed by planners.

As with Boolean models in general, the model can be displayed in a table - as in Figure 8.18. The left hand column contains (α, β, γ, δ, Π) inputs and the table gives (a, b, c, d, p) values for different (D, H, T) inputs. The underlined states in this table are the possible stable states. Equations (8.68) and (8.69) show that a state is stable if

$$a_{t+1} = a_t = \alpha_t \tag{8.75}$$

or, dropping the t subscripts,

$$a = \alpha \tag{8.76}$$

together with similar equations for (b,β), (c,γ), (d,δ) and (p,Π). The table shows that there are relatively few possible stable states. It is also a characteristic of Boolean models, as we would expect from their difference equation form, that cycles can exist.

The model described above has been applied to Brussels and gives a good, if broad, portrait of residential allocation in the city. The model obviously has a use at this broad level. Alternatively, it may be useful at more detailed scales of analysis which could be achieved if larger numbers of population classes were defined.

pabcd	D=0 H,T 0 0	0 1	1 0	1 1	D=1 H,T 0 0	0 1	1 0	1 1
00000	00110	00100	11110	11100	00110	00100	01110	01100
00001	00001	00000	00001	00000	00001	00000	00001	00000
00010	00110	00100	00110	00100	00010	00000	00010	00000
00011	00001	00000	00001	00000	00001	00000	00001	00000
00100	00110	00100	11110	11100	00110	00100	00110	00100
00101	00001	00000	00001	00000	00001	00000	00001	00000
00110	00110	00100	00110	00100	00010	00000	00010	00000
00111	00001	00000	00001	00000	00001	00000	00001	00000
01000	10110	10100	11110	11100	10110	10100	11110	11100
01001	00001	00000	00001	00000	00001	00000	00001	00000
01010	00110	00100	01110	01100	00010	00000	01010	01000
01011	00001	00000	00001	00000	00001	00000	00001	00000
01100	10110	10100	11110	11100	10110	10100	10110	10100
01101	00001	00000	00001	00000	00001	00000	00001	00000
01110	00110	00100	01110	01100	00010	00000	00010	00000
01111	00001	00000	00001	00000	00001	00000	00001	00000
10000	00100	00000	11100	11000	00100	00000	01100	01000
10001	00000	00000	00000	00000	00000	00000	00000	00000
10010	00100	00000	00100	00000	00000	00000	00000	00000
10011	00000	00000	00000	00000	00000	00000	00000	00000
10100	00100	00000	11100	11000	00100	00000	00100	00000
10101	00000	00000	00000	00000	00000	00000	00000	00000
10110	00100	00000	00100	00000	00000	00000	00000	00000
10111	00000	00000	00000	00000	00000	00000	00000	00000
11000	10100	10100	11100	11100	10100	10100	11100	11100
11001	00000	00000	00000	00000	00000	00000	00000	00000
11010	00100	00100	01100	01100	00000	00000	01000	01000
11011	00000	00000	00000	00000	00000	00000	00000	00000
11100	10100	10100	11100	11100	10100	10100	10100	10100
11101	00000	00000	00000	00000	00000	00000	00000	00000
11110	00100	00100	01100	01100	00000	00000	00000	00000
11111	00000	00000	00000	00000	00000	00000	00000	00000

Input variable :

D : population density
H : neighbourhood quality,
T : home-to-work travel time or cost

Internal variables :

π : housing price
α : high income residents
β : middle income residents
γ : autochthonal low income residents
δ : foreign low income residents

Potential demands :

p : variation of the housing price
migration of :
a : high income residents
b : middle income residents
c : autochthonal low income residents
d : foreign low income residents

Nature of the states

00000 : unstable
00000 : stable

Figure 8.18 Boolean state table for an urban system

Chapter 8

8.6.3 Difference equations and shopping centres

We now apply directly May's methods described in Section 8.4.2. Consider a set of shopping centres defined, as in Chapter 5, across a set of discrete zones. Let the size of the centre in zone j be W_j. Then, if D_j is the revenue potentially attracted to j, a suitable differential equation for the growth of W_j is, as we saw in Chapter 5:

$$\dot{W}_j = \varepsilon(D_j - kW_j) \tag{8.77}$$

for suitable constants ε and k. The difference equation form which suggests itself is

$$W_{jt+1} - W_{jt} = \varepsilon(D_j - kW_{jt}) \tag{8.78}$$

(where, without loss of generality, the time period is taken as 1, or as a factor emerged into ε). This can be written

$$W_{jt+1} = \varepsilon D_j + (1 - k)W_{jt} \tag{8.79}$$

Although this is a linear first order equation, and therefore does not have the interesting bifurcation properties of May's examples, we can apply his methods. Equation (8.79) expresses a linear relationship between W_{jt+1} and W_{jt} and an equilibrium point will be at the intersection of this line and the 45^0 line:

$$W_{jt+1} = W_{jt} \tag{8.80}$$

Various examples are shown in Figure 8.19. In relation to stability of equilibrium, the same argument applies as before: the slope of the 'curve' is now of course the gradient of the line and if this is between ± 1, then any intersection is stable - the argument of (8.47) and (8.48) above still holds. Four cases are distinguished on Figure 8.19: (a) and (c) are stable equilibrium points; in case (b) there is no equilibrium point with positive W_j; and in case (c), the equilibrium point is unstable. We can collect these results together in terms of the gradient of the line:

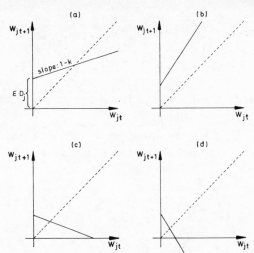

Figure 8.19 Retail centre size : W_{jt+1} **vs.** W_{jt} **for a linear difference equation II model**

(a) $\quad 0 < 1 - \varepsilon k < 1$ $\hspace{4cm}$ (8.81)

(b) $\quad 1 < 1 - \varepsilon k$ $\hspace{4.8cm}$ (8.82)

(c) $\quad -1 < 1 - \varepsilon k < 0$ $\hspace{3.5cm}$ (8.83)

(d) $\quad -1 > 1 - \varepsilon k$ $\hspace{4.3cm}$ (8.84)

The interesting and new feature about these relationships is that results about stability are related to general conditions involving two of the parameters in the model (8.79). Case (b) is immediately seen to be geographically nonsensical: it implies $\varepsilon k < 0$ when both of these parameters should be positive. (a) or (c) will hold provided the product εk is sufficiently small. Indeed, combining (8.81), (8.83) and (8.84), and assuming $\varepsilon k > 0$, the condition can be restated as

$$\varepsilon k < 2 \hspace{4cm} (8.85)$$

for stability, and

$$\varepsilon k > 2 \hspace{4cm} (8.86)$$

261

for instability. This also gives some clue as to the nature of the instabilities in difference equations. They arise because of the time lags involved in responding to a change. The greater the values of ε or k, the more rapid is the change from period to period and the more difficult it is to get back to equilibrium through feedback.

This analysis has been conducted as though D_j was fixed. In practice, of course, it is not and is given by

$$D_j = \sum_i S_{ij} \tag{8.87}$$

$$= \sum_i \frac{e_i P_i W_j^\alpha e^{-\beta c_{ij}}}{\sum_k W_k^\alpha e^{-\beta c_{ik}}} \tag{8.88}$$

since S_{ij}, the flow of revenue from residents of i to shops in j, is given in the usual way (see Chapter 5) by

$$S_{ij} = \frac{e_i P_i W_j^\alpha e^{-\beta c_{ij}}}{\sum_k W_k^\alpha e^{-\beta c_{ik}}} \tag{8.89}$$

e_i is per capita expenditure at i, P_i the population of i and c_{ij} the cost of travel from i to j. α and β are constants. In the analysis of equilibrium and stability in Chapter 5, the focus was on the stability of the equilibrium value, once it has been achieved. Equations (8.77) and (8.78) show that the equilibrium point is

$$W_j^* = D_j/K \tag{8.90}$$

It was shown in Chapter 5 that the stability of equilibrium depends on the values of the parameters like α, β and k. Here, we have seen that if D_j can be assumed constant, there is an additional condition (8.85). This can perhaps be interpreted as follows: if an equilibrium value of D_j is calculated using Harris and Wilson (1978) methods, as in Chapter 5, say as D_j^{equil}, then the ability to achieve a stable equilibrium in a simulation will require (8.85) to hold. Since it is a condition on parameters which are not j-dependent, this presumably means there will be difficulties in simulation in any cases where it is not satisfied. White (1977), for example, has reported simulations of this type which have not converged.

We now analyse an alternative form of model which is more directly analogous to May's example. The model given by (8.77) and (8.78) implies a steep rate of growth for W_j from a $W_j = 0$ starting point. This can be slowed down at the origin, but still bounded above, by adding a factor W_j by using a logistic form of growth (*cf.* Chapter 2 and Chapter 6). Equations (8.79) then becomes

$$\dot{W}_j = \varepsilon(D_j - kW_j)W_j \qquad (8.91)$$

This does not, of course, change the position of the equilibrium point which is still given by (8.90). We saw in Section 8.4.2 that there are at least two versions of difference equations which approximate logistic growth and we work with the one given by Equations (8.35) and (8.38). The obvious modification of Equation (8.91) gives

$$W_{jt+1} - W_{jt} = \varepsilon(D_j - kW_k)W_{jt} \qquad (8.92)$$

which can be written

$$W_{jt+1} = [(1 + \varepsilon D_j) - kW_{jt}]W_{jt} \qquad (8.93)$$

This is of the same form as Equation (8.35), and if we write

$$X_j = \frac{\varepsilon kW}{(1 + \varepsilon D_j)} \qquad (8.94)$$

then the equation takes the canonical form (8.38) with

$$a = 1 + \varepsilon D_j \qquad (8.95)$$

We can then immediately apply May's results on stability. Note that while D_j has the 'dimension' of money, Equation (8.92) shows ε to have the dimension of $(money)^{-1}$, and so εD_j is a dimensionless constant. The 'hump' of the curve in Figure 8.3 will be steeper for increasing values of either ε or D_j.

Chapter 8

A recapitulation of Section 8.4.2 shows that we require

$$1 < 1 + \varepsilon D_j < 3 \qquad (8.96)$$

for a stable single equilibrium point (using (8.40)), which is obviously

$$0 < \varepsilon D_j < 2 \qquad (8.97)$$

Clearly εD_j is always positive, but not necessarily less than 2. As it exceeds 2, then there is first a two-period cycle, then a four-period one, then some combination of three-cycles up to a chaotic regime which sets in at a = 3.8495, or εD_j = 2.8495. We should also recall that the system goes into divergent oscillations if a > 4, or $D_j > 3$.

As with model 1, D_j has been treated as a constant in this analysis. Again, a suitable first guess at it would be D_j^{equil} as predicted by the Harris and Wilson (1978) procedure described in Chapter 5. It is also more interesting in this case that the stability condition is j dependent, and even with D_j^{equil} it is dependent on the effects of any changes in other zones. This suggests the possibility of very complicated dynamic behaviour for a whole system which is evolving through the difference Equation (8.92).

The periodic, chaotic or divergent behaviour which results from εD_j exceeding 2 can arise in two ways which would need to be sorted out in particular empirical cases. First, since ε implicitly contains the time step length, it means that if this is too large there will be problems arising from such a (technical) choice. This means that special care will have to be taken if discrete simulations involving the logistic equations are used - as for example in the work of Allen and Sanglier (1979). Secondly, the instabilities arise in a real sense because of the implied feedback of the decision-maker which is represented in the discrete nature of the difference equation formulation, and it becomes a matter of empirical investigation as to whether these instabilities exist or not.

May (1976-A) has shown that very simple difference equations exhibit very complicated dynamic behaviour and he suggested a number of fields where the results were potentially applicable. We have shown in this subsection that they appear to have a direct application in geographical dynamics. It is perhaps a coincidence that the correspondence of Equations (8.92) with May's example is so exact, and of course this involves the restrictive condition that D_j should be treated as a constant. What will be even more interesting will be to explore the consequences of these kinds of bifurcation phenomena in more complicated economic models. For example, a retail model might be linked to a residential location model, as in Chapter 5, and this would, through the P_i's in Equation (8.88) have an impact on the D_j's. For particular values of ε, a 'jump' in D_j resulting from a P_i change could then lead, say, to new periodic behaviour in W_j.

It is also clear that, although the main argument has been cast here in terms of shopping centres, the methods and principles are more widely applicable to other urban structures. There is also beginning to be an extension of May's ideas to interacting populations in ecology - see for example Beddington, Free and Lawton (1975) on the investigation of dynamic complexity in prey-predator equations. There is much scope for numerical and empirical experiment and investigation.

Chapter 9

Concluding comments

The conclusions to be drawn from the studies presented in this book have been drawn in individual chapters. The purpose of this brief concluding chapter, therefore, is not to summarise such conclusions again, but to review the position which has been reached at a general level.

A striking feature is the diversity of models and approaches which has been generated in a relatively short time. Most of the references cited refer to the 1970's, and the majority of these to the last half of the decade. At the macro scale, the diversity of approach is perhaps, in the end, a weakness: there is a need for some integration; at the micro scale, there is diversity of types of example but as yet insufficiently concentrated effort to apply bifurcation theory in such fields as random utility theory. At the meso scale, there is promise of substantial advances in location theory, both in understanding the nature of 'comparative static' equilibrium models and in building fully dynamic models. The examples presented here have been rooted in spatial interaction models, but it should be emphasised that, while these often illustrate the point very well, the methods which have been exhibited could equally well be applied to alternative underpinning models.

So how far have we got? We can review progress at a general level by returning to Zeeman's six steps. The first is concerned with the analysis of the properties of equilibrium states. We have now seen enough evidence of the existence of 'catastrophic' change to know that the equilibrium properties of models will be examined much more carefully in the future. The next two steps relate to differential (or difference) equations:

the fast and slow dynamics.   We have also argued that there may
be a whole hierarchy of such levels in complex systems. We are
beginning to learn how to set up appropriate differential
equations, though perhaps not yet in an orderly and complete
system.   As with the analysis of equilibrium states, we have now
sufficient examples to know that bifurcation is interesting -
including transitions to periodic solutions as we saw at the end
of Chapter 8. More work is needed on the relative time scales
involved - the relaxation times - for different kinds of variable.

Zeeman's fourth step referred to building in feedback
effects, and here the systems analytical tradition of urban and
regional modelling has worked well: it is second nature for the
modeller to construct sets of simultaneous equations and to
attempt to include the most important feedbacks.   We may now
learn, however, from the new methods, that this often generates
new bifurcation phenomena.

The fifth step is concerned with noise and fluctuations. We
discussed the importance of this in the context of the work of
the Brussels school - particularly to chart development paths
for complex systems in cases where multiple steady-state solutions
are available and are conditioning the dynamics even when the
system never achieves equilibrium.

Finally, Zeeman discusses diffusion effects. Ultimately,
this leads to a concern with partial differential equations which
include space derivatives as well as time derivatives and
-elatively little progress has been made. However, the foundations have been laid by authors such as Hagerstrand (1967) and
it would be interesting to explore the bifurcation properties of
the corresponding differential equations (for example, those
presented by Webber and Joseph, 1978, 1979).

It has been useful to review progress in other disciplines
and to compare this with urban and regional studies. Two main
conclusions have perhaps emerged. First, it is still possible
for the geographer or social scientist to learn directly from
work in other disciplines - not by using analogies as such, but

by building on experience elsewhere of the applications of
certain kinds of methods. Secondly, the position of the geographer and social scientist compares favourably with other
disciplines in that the nature and difficulty of the problems
being studied are now demonstrably of the same order.

What next? There is immense scope for further development
of dynamical theory in various directions along the lines outlined in this book; and no doubt along other lines yet to be
discovered. The methods could be applied to alternative models
of the same systems studied here, or to different systems. There
is also the possibility of building an additional perspective for
the use of models in urban and regional planning, by focussing
on stability, criticality and bifurcation rather than simply the
more traditional conditional forecasting. Ideally, both
analytical and planning work should be carried forward empirically. This is likely to prove difficult because of the need
for time series data. However, only the development of theory
will encourage the appropriate data collection. In the meantime,
it is possible to get new insights for both theoretical and practical purposes by numerical experiments: running the models
using as much real data as possible, supplemented by realistic
guesses where it is not available. This adds some flesh to
the theoretical bones and provides a further stimulus to possible
empirical work.

APPENDIX 1

MATHEMATICAL PROGRAMMING FORMULATIONS OF THE MAIN MODELS

A1.1 Introduction

We indicated in Subsection 5.2.7 that there are various possible theoretical perspectives which form a basis for the main models used. They offer different kinds of insights, but the arguments of the main text are not affected by whichever one is chosen. Below, we outline in turn: entropy-maximising methods (A1.2); consumers' surplus and embedding theorem (A1.3, A1.4); accessibility maximisation (A1.5); random utility theory and group surplus (A1.6); applications to residential location (A1.7); and finally, applications to the Lowry model (A1.8).

A1.2 Entropy maximising and the shopping model

Because there are several possible interpretations of the concept of 'entropy' there are many ways to produce entropy maximising models. There is elementary discussion of this in Wilson (1970) and a much more thorough one, drawing on the concepts of information theory, by Webber (1980). The author's own style has always been to work by analogy with statistical mechanics; or to be more precise, to use the method (which is a high level method which can be applied in a number of disciplines) in the same way in geography as it has been applied in physics. Put this way, there should be no return to the old argument about whether a social science application in some sense *relies* on the analogy with physics. The 'analogy' is a consequence of the commonality of method. Thus, with these qualifications, the methods of (classical) statistical mechanics are used in this subsection.

It is possibly worthwhile trying to understand the key to the method for those, perhaps from other disciplines, who have not seen it before and so a largely conceptual presentation is made here and the reader is referred to Wilson (1970) or

271

Appendix 1

Wilson (1977-B) for a more detailed account. We use the retailing system as an example. The essence of the method is to describe the system of interest at three different levels of resolution: macro, meso and micro.

At the macro level, we recognise only the aggregate structural features of the system: that all consumers' cash is spent; that consumers derive a certain average (or more conveniently, but equivalently, total) benefit from shopping centres of different sizes; and that consumers incur travel costs which have an observed total (or again, alternatively average). These various totals are taken as $e_i P_i$ for each zone i, $\overline{\log W}$ and C respectively. The only awkward looking term is that proportional to the average value of $\log W_j$. We will see later that this can be related to the interpretation of the benefits of centre size: they grow with size (or attractiveness) but less than linearly, a notion which was introduced informally in Section 5.2.6 above.

At the meso scale, the main variables are the flow terms S_{ij}. It is clear that there are a large number of possible S_{ij}'s which are compatible with the macro-variables already defined. The constraints imposed from the macro scale onto the meso-scale variables can be written

$$\sum_j S_{ij} = e_i P_i \tag{A1.1}$$

$$\sum_{ij} S_{ij} \log W_j = \overline{\log W} \tag{A1.2}$$

$$\sum_{ij} S_{ij} c_{ij} = C \tag{A1.3}$$

At the micro scale, it can be assumed that, in principle, the state description consists of an account of the residences and associated shopping centres of each *individual* in the system. If the basic frame of reference is thought of as an origin-destination table, then since as at the next higher scale, there are many micro states which can give rise to any meso state, the

272

numbers of states at the various levels can be seen to be related in a hierarchical manner as shown in Figure A1.1.

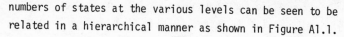

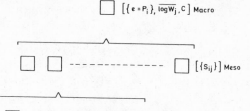

Figure A1.1 Hierarchical relationships between micro, meso and macro state descriptions

The essence of the method is to say that it is the meso scale we are most interested in and that micro states, subject to our turning up any more constraining information than already exists at the macro scale, *are all equally probable.* *Thus, we can find the most probable meso state, $\{S_{ij}\}$, by counting the number of micro states associated with each such meso state and selecting the $\{S_{ij}\}$ with the greatest number of micro states associated with it.*

This principle can be expressed mathematically after a number of basic steps. The number of micro states which can give rise to $\{S_{ij}\}$ can be shown to be (using the theory of permutations and combinations):

$$\frac{S!}{\prod_{ij} S_{ij}!} \quad (\text{where } S = \sum_{ij} S_{ij}) \tag{A1.4}$$

For convenience, it is the logarithm of this which is maximised:

$$Z = \log \left(\frac{S!}{\prod_{ij} S_{ij}!} \right) = \log S! - \sum_{ij} \log S_{ij}! \tag{A1.5}$$

subject to the constraints (A1.1)-(A1.3).

The solution comes out as

$$S_{ij} = e^{-\lambda_i - \alpha \log W_j - \beta c_{ij}} \tag{A1.6}$$

273

Appendix 1

where λ_i, α and β are the Lagrangian multipliers associated with constraints (A1.1)-(A1.3) respectively. If we define A_i such that

$$A_i e_i P_i = e^{-\lambda_i} \tag{A1.7}$$

and then calculate it by substituting in the Equation (A1.1) then

$$A_i = 1/\sum_j W_j e^{-\beta c_{ij}} \tag{A1.8}$$

which we immediately recognise as Equation (5.2). Here, of course, we have also observed that

$$e^{\alpha \log W_j} = W_j^{\alpha} \tag{A1.9}$$

so that Equation (A1.6) can be written in the form

$$S_{ij} = A_i e_i P_i W_j^{\alpha} e^{-\beta c_{ij}} \tag{A1.10}$$

which is simply Equation (5.1).

The method is known as an entropy-maximising method because the term Z in Equation (A1.5) has a family resemblance to the entropy term which occurs in physics. It is, however, less mystifying to think of the term as a measure of probability (since it is a count of micro states which we have assumed to be proportional to probability) so that the method can be thought of as probability maximising. Another way of doing the mathematics (known as the Darwin-Fowler method, and which produces the same answers) shows that the method can be viewed as a statistical averaging method: if some kind of average is taken over all possible micro states (subject only to the macro-level constraints), then the meso-state can be determined, and this produces the same spatial interaction model as that derived above. (This can also be seen intuitively from Figure A1.1.)

We could in theory seek the values of the Lagrangian multipliers α and β by substituting from (A1.6) into Equations (A1.2) and (A1.3) respectively. However, those equations are then non-linear in these multipliers and so it is

more convenient to leave them explicitly in the model equations
and to use the constraint equations for numerical solution as
part of the calibration process when data is available.   The
entropy-maximising formulation, however, does give further
insights into the nature of these model parameters by identifying them as Lagrangian multipliers.   They can then be interpreted
as shadow prices in relation to the constraint equations.   Note
also that there is a one-to-one relationship between the parameters and the values of $\overline{\log W}$ and C which occur on the right-
hand sides of the constraint equations.

For later convenience, we note that the method used to
solve the non-linear programming problem given by objective
function (A1.4) and constraints (A1.1)-(A1.3) is to form a
Lagrangian, L, given by

$$L = \log S! - \sum_{ij} \log S_{ij}! + \sum_i \lambda_i (e_i P_i - \sum_j S_{ij})$$

$$+ \ \alpha(\sum_{ij} S_{ij} \log W_j - \overline{\log W}) + \beta(C - \sum_{ij} T_{ij} c_{ij}) \tag{A1.11}$$

so that the problem can then be treated as an unconstrained one
with respect to the variables S_{ij} and the multipliers λ_i, α and
β.   We should also note that the trick for actually solving
the equations involves the use of Stirling's theorem (or, more
effectively, the alternative 'information theory' definition of
entropy) which enables the Lagrangian to be written in the form

$$L = - \sum_{ij} S_{ij} \log S_{ij} + \sum_i \lambda_i (e_i P_i - \sum_j S_{ij})$$

$$+ \ \alpha(\sum_{ij} S_{ij} \log W_j - \overline{\log W}) + \beta(C - \sum_{ij} T_{ij} c_{ij}) \tag{A1.12}$$

A1.3 Consumers′surplus

By noting the relationship in (A1.9) and with a little more
manipulation, we can write the basic aggregate model as

$$S_{ij} = A_i e_i P_i e^{\beta(\frac{\alpha}{\beta} \log W_j - c_{ij})} \tag{A1.13}$$

Appendix 1

which confirms our earlier interpretation in 5.2.6 of $\frac{\alpha}{\beta} \log W_j$ as a measure (in the same units as c_{ij}, say money) of the benefits of size, W_j. Thus, the expression

$$u_{ij} = \frac{\alpha}{\beta} \log W_j - c_{ij} \qquad (A1.14)$$

is the net benefit of being in the (i,j) state.

We can then usefulIly introduce the concept of consumers' surplus. Consider first that centre sizes are given and fixed, but that changes are possible in the c_{ij}'s. Then a plot of one S_{ij} against one c_{ij} is shown in Figure A1.2. It is assumed that all other variables are constant. The plot is assumed,

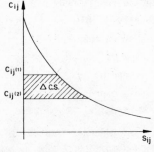

Figure A1.2 Change in consumers' surplus

notionally, to take the conventional form of a backward-sloping demand curve. It is customary to look at the change in consumers' surplus for a change in costs. Then the shaded area above the line of travel cost $c_{ij}^{(1)}$ can be interpreted as consumers' surplus in an obvious way: those consumers were prepared to pay more (and an amount given by that area) but do not have to. If the travel cost is reduced to $c_{ij}^{(2)}$, then the diagonally-shaded area represents the change in consumers' surplus.

This argument can easily be extended to work with benefits u_{ij} instead of simply travel costs c_{ij}, though if this is done, the argument carries the implication that W_j can vary as well as c_{ij}. We consider this argument in two stages: first, in the rest of this subsection that c_{ij} alone is varying, and then in the next that both W_j and c_{ij} can vary.

276

By combining certain constraints into the objective function via Lagrangian multipliers, but not the trip end constraints, the mathematical programming model of the previous section can be written:

$$\underset{\{S_{ij}\}}{\text{Max}} \quad Z = -\frac{1}{\beta} \sum_{ij} S_{ij} \log S_{ij} + \sum_{ij} S_{ij} \left(\frac{\alpha}{\beta} \log W_j - c_{ij}\right) \tag{A1.15}$$

subject to

$$\sum_j S_{ij} = e_i P_i \tag{A1.16}$$

where it should be noted that we have also divided through the terms in the objective function by β for reasons which will become apparent shortly.

It can then be shown that Z in (A1.15) can be interpreted as a measure of consumers' surplus for the shopping model problem. The significance of the dividing through by β is that the units of the whole expression are then the units of c_{ij}, say money, and in particular, dividing the entropy term by β converts it into money units. In this case of course, all the c_{ij}'s can be varying simultaneously and the expression is constructed by integrating each of the S_{ij} terms and summing to get a total consumers' surplus for a set of c_{ij}'s - the integration being from infinity to c_{ij} - *cf*. Figure A1.3, where we have reversed the axes from Figure A1.2. Thus, the entropy-maximising problem can be interpreted formally, as a consumers'

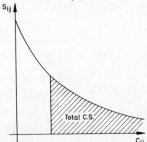

Figure A1.3 Total consumers' surplus, axes reversed

surplus maximisation problem. This is formal in the sense that we have taken the 'demand' function, the expression for S_{ij}

277

Appendix 1

resulting from the entropy-maximisation procedure, and by
integration found the expression for consumers' surplus, and
then said that if you want to interpret it the other way round,
you can. However, such interpretations are best based on random
utility theory, and this approach will be developed further
below in Section A1.6.

Finally, we note that with the objective function in the
form (A1.15) we can see what happens as $\beta \rightarrow \infty$. The entropy
term is knocked out of the objective function and, since we are
assuming fixed W_j's, we can ignore those and the result will be
to assign consumers to the nearest shopping centre. Because
the model is singly constrained, this is a relatively trivial
version of a result which is more interesting in the doubly-
constrained case (Evans, 1973, Wilson and Senior, 1974).

A1.4 Embedding: optimum centre size and location

The next step in the argument is to consider similar
programming problems to those of the previous subsection but
within which W_j can vary as well as S_{ij}. The problem which is
then being considered is to find the optimum size and locations
of shopping centres - and essentially, in the first instance,
for fixed c_{ij} so that in any consumers' surplus measures relate
to u_{ij} and to W_j changes within it.

It is useful to begin with an old formulation of this problem which offers new insights into what can be achieved
through certain kinds of mathematical programming formulations.
Suppose that the objective function can be written in the form

$$Z = f(\underline{W}) + h(\underline{S},\underline{W}) \qquad \qquad (A1.17)$$

(where $\underline{W}$ and $\underline{S}$ are a vector and matrix respectively representing
$\{W_j\}$ and $\{S_{ij}\}$ respectively). For later convenience, this
objective function has been shown as a sum of functions of $\underline{W}$
alone and $\underline{W}$ and $\underline{S}$ together. What we would like to do in principle is to choose the functions f and h to represent some
appropriate objectives for a particular system - whether profit-
maximising, government acting on behalf of consumers or whatever -

278

and maximise Z subject to constraints which represent the
assumption that consumers will continue to behave according to
the spatial interaction model.   Thus, the problem of the optimisation of size and location of shopping centres could be said
to be: maximise Z subject to constraints (5.1) and (5.2).
Especially because of the A_i terms, these constraints are highly
non-linear. The first attempt to formulate this optimisation
problem involved taking A as the sum of benefits as measured by
u_{ij} in Equations (A1.14)

$$Z = \sum_{ij} S_{ij} u_{ij} = \sum_{ij} S_{ij}(\frac{\alpha}{\beta} \log W_j - c_{ij}) \qquad (A1.18)$$

Since $\{W_j\}$ now vary, we need to constrain them in total by

$$\sum_j W_j = W \qquad (A1.19)$$

The problem is now to maximise Z over $\underline{S}$ and $\underline{W}$ subject to the
model constraints (5.1) and (5.2) together with (A1.19).
Unfortunately, because of the non-linearities in both the objective function and the constraints, there is no standard way of
solving this problem either analytically or numerically. However,
it can be shown that the problem given by Equations (A1.18),
(A1.19), (5.1) and (5.2) is equivalent to one which can be
written in the form

$$Z = -\frac{1}{\beta} \sum_{ij} S_{ij} \log S_{ij} + \sum_{ij} S_{ij}(\frac{\alpha}{\beta} \log W_j - c_{ij}) \qquad (A1.20)$$

subject to

$$\sum_j S_{ij} = e_i P_i \qquad (A1.21)$$

and

$$\sum_j W_j = W \qquad (A1.22)$$

*and that the resulting S_{ij} system then satisfies the shopping
model Equations (5.1) and (5.2).* In effect, what has happened
is that the shopping model equations used as constraints have
been decomposed into their entropy-maximising (or consumers'

Appendix 1

surplus maximising) form - the non-linear parts have been
absorbed into the objective function and only linear constraints
retained. In effect, one mathematical programming problem -
the $\underline{S}$ problem - has been absorbed into another one - the $\underline{W}$ one,
or we can say that one has been *embedded* into the other. The
benefit of embedding is that it suggests new ways of writing the
programme, as we have seen, which are then computationally more
convenient: a programme such as (A1.20)-(A1.22) which has a
non-linear objective function but wholly linear constraints *can*
be solved, computationally at least.

This embedding theorem can be expressed more generally
(Coelho and Wilson, 1977; Coelho, Williams and Wilson, 1978):
for Z in the form of Equation (A1.7) then if f is any concave
function and any convex constraints are added, then provided h
takes the form

$$h = - \sum_{ij} S_{ij} (\log S_{ij} - 1 + \alpha \log W_j - \beta c_{ij}) \quad (A1.23)$$

then the S_{ij} variables always satisfy the shopping model equations. Note that this result does of course properly belong to
the section which introduces $\underline{W}$ variation as well as $\underline{S}$ variation
into the problem. This can be seen more clearly in the dis-
aggregated form of the embedding theorem which can be stated
formally as follows (using an obvious notation for person
classes, n): if $f(\underline{W})$ is a concave function, $g_k(\underline{W})$ are convex
functions defining constraints

$$g_k(\underline{W}) < b_k \quad (A1.24)$$

and that these equations, together with the constraints

$$\sum_j S_{ij}^n = e_i^n p_i^n \quad (A1.25)$$

form a non-empty feasible set, and $h(\underline{S},\underline{W})$ is defined as

$$h(\underline{S},\underline{W}) = - \sum_{ijn} \mu_n S_{ij}^n (\log S_{ij}^n - 1 + \alpha^n \log W_j - \beta^n c_{ij}) \quad (A1.26)$$

280

then the problem

$$\underset{(\underline{W},\underline{S})}{\text{Max}} \quad Z = f(\underline{W}) + h(\underline{S},\underline{W}) \tag{A1.27}$$

subject to the constraints (A1.24) and (A1.25) has a unique solution and S_{ij}^n satisfies:

$$S_{ij}^n = A_i^n e_i^n P_i^n W_j^{\alpha^n} e^{-\beta^n c_{ij}^n} \tag{A1.28}$$

where the μ_n in (A1.26) are arbitrary positive constants.

This leads us to note another interesting result: if we choose $\mu_n = 1$, then the programme is a standard disaggregated entropy maximising model. If we choose $\mu_n = \frac{1}{\beta_n}$, then the objective function can be shown to represent consumers' surplus. Thus, while the entropy-maximising and consumers' surplus maximising models are identical in the aggregate case, they differ in the disaggregate case, and the difference relates to the position of the parameters in the objective function. They have to divide individually into the entropy-like terms into order to convert these into money units so that the whole expression can be interpreted as consumers' surplus.

The demand functions so constructed also satisfy the Hotelling integrability conditions. With the entropy form, the β_n- parameters appear as Lagrangian multipliers on the cost terms and so play a different role. The resulting demand functions do not satisfy the integrability conditions. In the case where all the different β_n's take the same value, then the results are of course identical (and the results are approximately the same if the β_n's are approximately equal). As noted earlier, we can gain more insight into the nature of consumers' surplus in the disaggregated case from the random utility approach which we take up in Section A1.5 below.

Appendix 1

A1.5 Accessibility maximising: Leonardi's formulation

Leonardi (1973, 1975, 1977) has posed an interesting mathematical programming problem which turns out (at least for the aggregate case) to be equivalent to the other formulations above. His problem is:

$$\underset{\{W_j\}}{\text{Max}} \quad Z = \sum_i e_i P_i \log \sum_j W_j e^{-\beta c_{ij}} \qquad (A1.29)$$

subject to

$$\sum_j W_j = W \qquad (A1.30)$$

The term

$$X_i = \sum_j W_j e^{-\beta c_{ij}} \qquad (A1.31)$$

has long been interpreted as a good measure of accessibility (Hansen, 1959) so that the objective function (A1.29) can be interpreted as maximising what may be called log accessibility. The problem in this form has as an optimality condition

$$\frac{\partial Z}{\partial W_j} = \lambda \qquad (A1.32)$$

where λ is the Lagrangian multiplier associated with (A1.31). This can be shown to imply (provided the S_{ij}'s are assumed to satisfy their basic model Equations (5.1) and (5.2)) that

$$\sum_i S_{ij} = \gamma W_j \qquad (A1.33)$$

which is obviously similar to the 'balancing' condition we discussed in Section 5.1 of Chapter 5 above, as one of the possible hypotheses for determining the W_j's. But, given that Leonardi's formulation can be shown to be equivalent to the others; we have the rather curious result that hypotheses based on producers' behaviour give the same results, for W_j, as those based on consumers' behaviour. This is probably due to the rather tight constraints which are imposed in this case. (We should also

282

note in passing that the basis of the proof of the equivalence
of Leonardi's formulation with the others turns on the duals of
the others: essentially Leonardi is taking the dual form and
eliminating any explicit appearance of the S_{ij} variables -
though, as we mentioned, these are needed to derive the
appropriate optimality conditions.)

A1.6 Random utility theory and group surplus

So far, the theoretical approaches presented have been
explicitly at the meso scale. In some ways, this is perhaps
appropriate for meso-scale phenomena, and the entropy-
maximising approach still has an intuitive appeal on this basis.
The consumers' surplus argument presented above essentially
relied on the observation that when the demand function was
integrated, consumers' surplus could be identified and then the
converse argument applied: that the demand function could be
seen as arising out of the maximisation of consumers' surplus.
In both cases, it can be argued that it is better to seek a
micro-scale theoretical approach based on some hypotheses about
consumers' behaviour. This has always turned out to be a
difficult problem because of the aggregation problem: how to
get meso-scale results from micro theory. An approach which
is now available to solve this problem is that of random utility
theory. The barest sketch will be given here, as it applies to
the shopping model. For further details, see, for example,
Williams (1977), Coelho (1977), Coelho and Williams (1977).

The basic idea is that each consumer maximises his utility
but that this is subject to some random variation - either to
account for unmeasured differences in preferences, lack of
information, or whatever. Suppose that the net benefit a consumer gets from shopping in zone j (if he is a resident of
zone i) is

$$u_{ij} = u_j - c_{ij} + \varepsilon_j \qquad (A1.34)$$

where u_j is the mean benefit of shopping in j (in other words a
more general version of the term $\frac{\alpha}{\beta} \log W_j$ which we have had up

283

Appendix 1

to now). c_{ij} is the usual transport cost and ε_j is a random variable which is assumed to be Weibull distributed about zero. Then, the probability of a resident of zone i choosing to shop in zone j is

$$p_{ij} = Pr(u_{ij} > u_{ij'}, \text{ all } j' \neq j) \tag{A1.35}$$

On the basis of the assumption of the Weibull distribution for the c's, this can be calculated and it gives the result:

$$p_{ij} = \frac{e^{\beta(u_j - c_{ij})}}{\sum_j e^{\beta(u_j - c_{ij})}} \tag{A1.36}$$

Then, since

$$S_{ij} = e_i P_i p_{ij} \tag{A1.37}$$

and we can take

$$U_j = \frac{\alpha}{\beta} \log W_j \tag{A1.38}$$

for present purposes, then substituting from Equation (A1.26) for p_{ij} and (A1.28) for U_j, we get

$$S_{ij} = A_i e_i P_i W_j^\alpha e^{-\beta c_{ij}} \tag{A1.39}$$

which is the standard shopping model formula yet again.

However, we can now take the argument a step further: it can be shown that the average consumers' surplus for a resident of zone i is:

$$Z_i = -\frac{1}{\beta} \sum_j p_{ij} \log p_i + \sum_j p_{ij}(u_j - c_{ij}) \tag{A1.40}$$

and if we multiply by $e_i P_i$ and sum over i we can see that the total consumers' surplus is:

$$Z = -\frac{1}{\beta} \sum_{ij} S_{ij} \log S_{ij} + \sum_{ij} S_{ij}(\frac{\alpha}{\beta} \log W_j - c_{ij}) \tag{A1.41}$$

and that if this is maximised subject to the usual constraint

$$\sum_j S_{ij} = e_i P_i \tag{A1.42}$$

284

then we have our usual form of the mathematical programming model for S_{ij}, but with a different derivation. Note the position of the β- parameter in the consumers' surplus term. It is a straightforward matter to add W variation and the additional constraint on the sum of the W_j's and also to produce the sort of disaggregated model we presented in the discussion of embedding above.

A1.7 Application to residential location models

It will already be clear from the discussion in Section 5.1 that the structure of the basic residential location model is essentially the same as that of the shopping model. Both are singly constrained spatial interaction models, the shopping model being usually considered as production constrained, the residential location model as attraction constrained. It is therefore a straightforward matter to apply all the results of the previous subsections, which have been presented for the shopping model, to the residential location model. This is left as an exercise to be taken up as required - and is in part carried out in the next subsection on the Lowry model in relation to its residential location component.

A1.8 Mathematical programming versions of the Lowry model

The essence of the original Lowry model, and many others of the same type, was outlined in Section 5.2.4. There are a great many ways in which the Lowry model can be represented as a single mathematical programming model - pioneered by Coelho and Williams (1977). These are discussed in some detail in the paper by Macgill and Wilson (1979). Here, we shall simply look at the basic principles for constructing such models and choose one aggregate example as an illustration, and then develop a suitable disaggregated version for Chapter 6, in the section on central place theory below.

The first obvious step to produce a single mathematical programme for the Lowry model is to combine the separate entropy models for residential location and service use, and the corresponding set of constraint equations - in the first case, as in the retailing example, taking the stock variables (now P_i and E_j) as fixed.   There is then a by-now familiar generalisation which allows these stock variables to be determined within that mathematical programme - so that the pattern which results is really determined only by the given distribution of basic employment, the c_{ij}-matrix and any intrinsic attractiveness factors. One interesting result to emerge is that the form of the constraints for the linked models, when combined into a single programming model, creates a residential location pattern which is dependent both on access to services as well as the more usual access to jobs, and so this makes the interacting field concept introduced in Section 5.2.4 appear more complicated. It is also possible to continue this sequence of argument in the direction of random utility models: we take one such model, that developed by Coelho and Williams (more precisely a consumers' surplus model which results from such an analysis) as our example.

In this particular example, we introduce, following Coelho and Williams (1977), a set of coefficients which illustrate how to get the units right for the set of interaction models involved in the Lowry model, and in this sense refine our earlier presentation of the component spatial interaction models.   We begin, therefore by writing down the accounting equations. Let the interaction variables $\hat{T}_{ij}$ and $\hat{S}_{ij}$ define the number of jobs allocated to population at i from zone j, and the number of service jobs created at j by residents of zone i.   In a sense, these are our basic units, but by defining coefficients η and ρ we can convert these to trips:

$$\hat{T}_{ij} = \eta T_{ij} \qquad (A1.43)$$

$$\hat{S}_{ij} = \rho S_{ij} \qquad (A1.44)$$

and work with these variables. Then, if α is an inverse activity rate and σ_i is the number of service jobs generated by the population at i (like our earlier e_i, but now in jobs units instead of money units), then we have the relations

$$\sum_j \hat{T}_{ij} = \frac{1}{\alpha} P_i \qquad (A1.45)$$

$$\sum_j \hat{S}_{ij} = \sigma_i P_i \qquad (A1.46)$$

which can be written in terms of the T_{ij} and S_{ij} variables as

$$\sum_j T_{ij} = \frac{\eta}{\alpha} P_i \qquad (A1.47)$$

$$\sum_j S_{ij} = \sigma_i \rho P_i \qquad (A1.48)$$

using (A1.43) and (A1.44) respectively.

We also have constraints in the employment end of the trips:

$$\sum_i T_{ij} = \eta E_j \qquad (A1.49)$$

$$\sum_i S_{ij} = \rho E_j^R \qquad (A1.50)$$

together with the equation (given earlier in Chapter 5 as (5.23) and repeated here for convenience) which relates total employment to basic and retail (or service) employment:

$$E_j = E_j^B + E_j^R \qquad (A1.51)$$

The only remaining step is to write down an objective function which contains both kinds of trip terms, and this is done in this example using the group consumers' surplus criterion based on random utility theory (so the β-weights appear on the utility terms). We also, to illustrate yet another way of writing these things, follow Coelho and Williams in including the attractiveness terms within the entropy function rather than as a separate 'benefit' term. Formally, this does not make any difference in this particular model but there are possible differences in interpretation: in the form given below, the attractiveness terms can be taken as representing prior

Appendix 1

probabilities with the entropy term taking the Kullback form. (See the references for further details on this.) It can be argued that it is important to deal with intrinsic properties of zones, such as size in this way, but this turns on whether service centres are to be interpreted as directly related to zone size or as 'points' within a zone. In the residential case, it may be more reasonable to associate attractiveness partly with size. However, this debate takes us beyond what we need for present purposes and is not pursued further here. The objective function is therefore:

$$\max_{\{T_{ij}, S_{ij}\}} Z = -\frac{1}{\mu} \sum_{ij} T_{ij} (\log \frac{T_{ij}}{W_i^{res}} - 1) - \sum_{ij} T_{ij} c_{ij}$$

$$-\frac{1}{\beta} \sum_{ij} S_{ij} (\log \frac{S_{ij}}{W_j}) - \sum_{ij} S_{ij} c_{ij} \qquad (A1.52)$$

Thus, the problem is to maximise Z with respect to T_{ij} and S_{ij}. E_j^R, E_j and P_i are determined by the constraints (A1.47)-(A1.51) which must be satisfied simultaneously. In their formulation, $\{W_i^{res}\}$ and $\{W_j\}$ are being used more passively than hitherto as intrinsic and exogenous attractiveness weights.

A somewhat more elgant form of the model, and one which gives us new insights, can be obtained by eliminating the variables P_i, E_j and E_j^R from Equations (A1.47)-(A1.51) which gives as an alternative constraint set

$$\sum_j T_{ij} - \lambda_{1i} \sum_j S_{ij} = 0 \qquad (A1.53)$$

$$\sum_i T_{ij} - \lambda_2 \sum_i S_{ij} = \eta E_j^B \qquad (A1.54)$$

where λ_{1i} and λ_2 are defined by

$$\lambda_{1i} = \eta/\rho\alpha\sigma_i \qquad (A1.55)$$

$$\lambda_2 = \eta/\rho \qquad (A1.56)$$

When the model is used in this form, the variables which have been eliminated can always of course then be calculated from the constraint equations in their original form.

The form of the constraints (A1.53) and (A1.54) show that there is a close relationship between T_{ij} and S_{ij} variables in this form of model.

These formulations are used in Chapter 6 when looking at dynamical forms of Lowry model.

APPENDIX 2

SOME ALTERNATIVE LAGRANGIAN FORMULATIONS

We noted in Section 5.3.2 that when a problem could be formulated in Lagrangian form, then, for a maximisation problem, the first idea for a possible dynamical equation is taken as

$$\dot{W}_j = \frac{\partial L}{\partial W_j} \qquad (A2.1)$$

The Lagrangian for the retail problem is (repeating (5.41) for convenience):

$$L = -\sum_{ij} S_{ij} \log S_{ij} + \sum_i P_i(e_i P_i - \sum_j S_{ij}) + \alpha(\sum_{ij} S_{ij} \log W_j - \overline{\log W_j})$$

$$+ \beta(C - \sum_{ij} S_{ij} c_{ij}) + \gamma(W - \sum_j W_j) \qquad (A2.2)$$

It is easy to see that if this is differentiated with respect to W_j, we get

$$\frac{\partial L}{\partial W_j} = \frac{\alpha \sum_i S_{ij}}{W_j} - \gamma \qquad (2.3)$$

so that Equation (A2.1) takes the form:

$$\dot{W}_j = \frac{\alpha}{W_j} \left[\sum_i S_{ij} - \frac{\gamma}{\alpha} W_j \right] \qquad (A2.4)$$

This, of course, is not the same as the main equation we use, which is also repeated here for convenience:

$$\dot{W}_j = \varepsilon [\sum_i S_{ij} - kW_j] \qquad (A2.5)$$

The difference between Equations (A2.4) and (A2.5) lies not in the position of the equilibrium points, but in the form of take-off from $W_j = 0$, as we discussed for general forms of equation

291

Appendix 2

of this type in Chapter 2.[1] The difference, therefore, is not necessarily of any great significance.

   The remaining point to note, however, is that equations of the form (A2.5) can be derived from a Lagrangian. Consider the general differential equation

$$\dot{W}_j = \epsilon W_j^n (D_j - kW_j) \tag{A2.6}$$

(the form of which was originally introduced as Equation (2.12) in Chapter 2). This can easily be seen to arise from the Lagrangian

$$L = -\sum_{ij} S_{ij} \log S_{ij} + \sum_i \lambda_i (e_i P_i - \sum_j S_{ij}) + \alpha \sum_{ij} S_{ij} (W_j^{(n+1)} - \overline{W_j^{(n+1)}})$$

$$+ \beta (C - \sum_{ij} S_{ij} c_{ij}) + \gamma (W^{n+2} - \sum_j \overline{W_j^{n+2}}) \tag{A2.7}$$

However, now, we clearly have much greater difficulty interpreting the constraint term. The attractiveness factor constraint leading to this Lagrangian is

$$\sum_{ij} S_{ij} W_j^{n+1} = \overline{W_j^{n+1}} \tag{A2.8}$$

and the one constraining total space is

$$\sum_j W_j^{n+2} = \overline{W_j^{n+2}} \tag{A2.9}$$

(where we show appropriately defined averages on the right-hand side). The Lagrangian (A2.2) is the special case of (A2.7) for $n = -1$. The differential equation which arises from (A2.7) is

$$\dot{W}_j = \frac{\partial L}{\partial W_j} = \alpha(n+1)W_j^n S_{ij} - \gamma(n+2)W_j^{n+1} = 0 \tag{A2.10}$$

which is the same as (A2.6) provided that we take

$$\epsilon = \alpha(n+1) \tag{A2.11}$$

and

$$k = \frac{\gamma(n+2)}{\alpha(n+1)}, \tag{A2.12}$$

The effect of the constraint (A2.8) is to generate a new attractiveness factor so that the S_{ij}'s satisfy

$$S_{ij} = A_i e_i P_i e^{\alpha W_j^{n+1}} e^{-\beta c_{ij}} \qquad (A2.13)$$

Thus, for n = 0, we get an exponential form of attractiveness function, and for higher values of n, an exponential power function. This, at least, is plausible in principle. The constraint (A2.9) is much more difficult to interpret, since it is intended to be a constraint on total space and not on this quantity raised to a power. Even for n = 0, this is difficult to interpret. The remaining possibility, therefore, is to allow ourselves to be led to a new form of model by retaining the constraint Equation (A2.8) in order to get the W_j^n factor in the differential equation, but replacing the constraint (A2.9) by the more conventional constraint

$$\sum_j W_j = W \qquad (A2.14)$$

Some algebra shows that this generates a differential equation of the form

$$\dot{W}_j = \frac{\partial L}{\partial W_j} = \alpha(n+1)W_j^n \left(\sum_j S_{ij} - \frac{\gamma}{W_j^n} \right) \qquad (A2.15)$$

This is interesting because the term in the position of γ/W_j^n is usually interpreted as representing the cost of provision of facilities, and this now implies, for n > 0, the existence of scale economies. So, in an indirect way, we may have obtained a clue on how to build scale economies into the supply side dynamics.

Note

1.   Provided, of course, that we take k = γ/α and define a suitable constant ε - either by adding a factor to (A2.1) or by taking $\varepsilon = \alpha$.

APPENDIX 3

THE DERIVATIVES OF S_{ij}

If we substitute for A_i from Equation (5.2) into Equation (5.1), then S_{ij} can be written

$$S_{ij} = \frac{e_i P_i W_j^\alpha e^{-\beta c_{ij}}}{\sum_k W_k^\alpha e^{-\beta c_{ik}}} = \frac{e_i P_i W_j^\alpha e^{-\beta c_{ij}}}{X_i} \qquad (A3.1)$$

where for convenience, we define

$$X_i = \sum_k W_k^\alpha e^{-\beta c_{ik}} \qquad (A3.2)$$

Then

$$\frac{\partial S_{ij}}{\partial W_j} = \frac{e_i P_i \alpha W_j^{\alpha-1} e^{-\beta c_{ij}}}{X_i} - \frac{\alpha W_j^{\alpha-1} . e_i P_i W_j^\alpha e^{-\beta c_{ij}}}{X_i} \qquad (A3.3)$$

$$= \frac{\alpha e_i P_i W_j^{\alpha-1} e^{-\beta c_{ij}}}{X_i} \left[1 - \frac{W_j^\alpha e^{-\beta c_{ij}}}{X_i} \right] \qquad (A3.4)$$

$$= \frac{\alpha S_{ij}}{W_j} \left[1 - \frac{S_{ij}}{e_i P_i} \right] \qquad (A3.5)$$

by noting the form of the first factor and the last term in the brackets of (A3.4) and comparing them with (A3.1).

It is easiest to compute the second derivative by differentiating (A3.5) again and then using the result we have already obtained for the first derivative. This leads to the following sequence of calculations:

295

Appendix 3

$$\frac{\partial^2 S_{ij}}{\partial W_j^2} = \frac{\alpha}{W_j} \frac{\partial S_{ij}}{\partial W_j} - \frac{\alpha}{W_j^2} S_{ij} - \frac{\alpha}{W_j e_i P_i} .2S_{ij} \frac{\partial S_{ij}}{\partial W_j} + \frac{\alpha S_{ij}^2}{W_j^2 e_i P_i} \quad \text{(A3.6)}$$

$$= \frac{\alpha}{W_j} . \frac{\alpha S_{ij}}{W_j} (1 - \frac{S_{ij}}{e_i P_i}) - \frac{\alpha}{W_j^2} S_{ij} - \frac{2\alpha S_{ij}}{e_i P_i W_j} \frac{\alpha S_{ij}}{W_j} \times$$

$$(1 - \frac{S_{ij}}{e_i P_i}) + \frac{\alpha S_{ij}^2}{W_j^2 e_i P_i} \quad \text{(A3.7)}$$

$$= \frac{\alpha^2 S_{ij}}{W_j^2} - \frac{\alpha^2 S_{ij}^2}{W_j^2 e_i P_i} - \frac{\alpha}{W_j^2} S_{ij} - \frac{2\alpha^2 S_{ij}^2}{e_i P_i W_j^2} + \frac{2\alpha^2 S_{ij}^3}{e_i^2 P_i^2 W_j^2} + \frac{\alpha S_{ij}^2}{W_j^2 e_i P_i}$$

$$\text{(A3.8)}$$

$$= \frac{\alpha S_{ij}}{W_j} \left[(\alpha - 1) + (1 - 3\alpha) \frac{S_{ij}}{e_i P_i} + 2\alpha (\frac{S_{ij}}{e_i P_i})^2 \right] \quad \text{(A3.9)}$$

This last expression can be factored to give

$$\frac{\partial^2 S_{ij}}{\partial W_j^2} = \frac{2\alpha^2 S_{ij}}{W_j} (\frac{S_{ij}}{e_i P_i} - 1) \left[\frac{S_{ij}}{e_i P_i} - (\frac{\alpha - 1}{2\alpha}) \right] \text{(A3.10)}$$

APPENDIX 4

THE SHOPPING TRIP FLOW DERIVATIVES FOR THE DISAGGREGATED MODEL

In the previous Appendix, we showed how to calculate the derivatives of S_{ij} for the aggregate shopping model. Here, we consider the disaggregated model, using essentially the model of Section 5.4.6, but in a more general form, using some of the results of the model of Section 6.4.2 (but neglecting work zone origins).

We take the model in the form

$$S_{ij}^g = A_i^g e_i^g P_i \hat{W}_j^g e^{-\beta^g c_{ij}} \qquad (A4.1)$$

where

$$A_i^g = 1 / \sum_k \hat{W}_k^g e^{-\beta^g c_{ik}} \qquad (A4.2)$$

We can also usefully define

$$\chi_i^g = \sum_k \hat{W}_k^g e^{-\beta^g c_{ik}} \qquad (A4.3)$$

and rewrite equation (A4.1) as

$$S_{ij}^g = \frac{e_i^g P_i \hat{W}_j^g e^{-\beta c_{ij}}}{\chi_i^g} \qquad (A4.4)$$

The composite attractiveness factor is given by

$$\hat{W}_j^g = (\sum_{g'} W_j^{g'})^{\alpha_1^g} \prod_{g'} (W_j^{g'})^{\alpha_2^{g'g}} \qquad (A4.5)$$

The array $\alpha_2^{g'g}$ determined the influence of the provision of one particular good on the attractiveness of another. Obviously the most important term is the α_2^{gg} diagonal element and in the model of Section 5.4.6 that is the only one present. Clearly also,

Appendix 4

many elements of the array will be zero. However, this general form may be useful as a way of building some hierarchical structure into models. For example, if $\alpha_2^{g'\,g}$ was non-zero whenever g' was of a lower order than g, this would mean that $\hat{W}_j^g$ was zero whenever $W_j^{g'}$ was zero for a lower order g'. In other words, higher order goods in a centre could only survive if it was also possible for consumers to acquire lower order goods at the same time.

Then, differentiating (A4.4), we have

$$\frac{\partial S_{ij}}{\partial W_j^g} = e_i^g P_i e^{-\beta c_{ij}} \left[\frac{1}{X_i^g} \frac{\partial \hat{W}_j^g}{\partial W_j^g} - \frac{\hat{W}_j^g}{X_i^{g2}} \frac{\partial X_i^g}{\partial W_j^g} \right] \tag{A4.6}$$

$$= \frac{e_i^g P_i \hat{W}_j^g e^{-\beta c_{ij}}}{X_i^g} \left[\frac{1}{\hat{W}_j^g} \frac{\partial \hat{W}_j^g}{\partial W_j^g} - \frac{1}{X_i^g} \frac{\partial X_i^g}{\partial W_j^g} \right] \tag{A4.7}$$

$$= S_{ij}^g \left[\frac{1}{\hat{W}_j^g} \frac{\hat{W}_j^g}{\partial W_j^g} - \frac{1}{X_i^g} \frac{\partial X_i^g}{\partial W_j^g} \right] \tag{A4.8}$$

Differentiating (A4.3), we have

$$\frac{\partial X_i^g}{\partial W_j} = \sum_{j'} \frac{\partial \hat{W}_{j'}^g}{\partial W_j^g} e^{-\beta c_{ij'}} \tag{A4.9}$$

Note, however, from Equation (A4.5) that if we substitute j' for j in that equation, then no W_j^g terms remain, so that

$$\frac{\partial \hat{W}_j^g}{\partial W_{j'}^g} = 0 , \quad j' \neq j \tag{A4.10}$$

Hence

$$\frac{\partial X_i^g}{\partial W_j^g} = \frac{\partial \hat{W}_j^g}{\partial W_j} \cdot e^{-\beta c_{ij}} \tag{A4.11}$$

298

We can substitute this expression into Equation (A4.8) and get $\dfrac{\partial S_{ij}^g}{\partial W_j^g}$ in terms of $\dfrac{\partial \hat{W}_j^g}{\partial W_j^g}$ only:

$$\frac{\partial S_{ij}^g}{\partial W_j^g} = S_{ij}^g \left(\frac{1}{\hat{W}_j^g} \frac{\partial \hat{W}_j^g}{\partial W_j^g} - \frac{e^{-\beta c_{ij}}}{X_i^g} \frac{\partial \hat{W}_j^g}{\partial W_j^g} \right) \tag{A4.12}$$

$$= \frac{S_{ij}^g}{\hat{W}_j^g} \frac{\partial \hat{W}_j^g}{\partial W_j^g} \left(1 - \frac{e^{-\beta c_{ij}} \hat{W}_j^g}{X_i^g} \right) \tag{A4.13}$$

$$= \frac{S_{ij}^g}{\hat{W}_j^g} \frac{\partial \hat{W}_j^g}{\partial W_j^g} \left(1 - \frac{S_{ij}}{e_i P_i} \right) \tag{A4.14}$$

It now only remains to calculate $\dfrac{\partial \hat{W}_j^g}{\partial W_j^g}$. Differentiating Equation (A4.5) we get

$$\frac{\partial \hat{W}_j^g}{\partial W_j^g} = \alpha_1^g \left(\sum_{g'} W_j^{g'} \right)^{\alpha_1^g - 1} \prod_{g'} (W_j^{g'})^{\alpha_2^{g'g}} + \left(\sum_g W_j^{g'} \right)^{\alpha_1^g} \times$$

$$\prod_{g' \neq g} (W_j^{g'})^{\alpha_2^{g'g}} {}_{\alpha_2}^{gg} W_j^g (\alpha_2^{gg} - 1) \tag{A4.15}$$

$$= \left(\sum_{g'} W_j^{g'} \right)^{\alpha_1^g} \prod_{g'} (W_j^{g'})^{\alpha_2^{g'g}} \left[\frac{\alpha_1^g}{\left(\sum_{g'} W_j^{g'} \right)} + \frac{\alpha_2^{gg}}{W_j^g} \right] \tag{A4.16}$$

$$= \hat{W}_j^g \left[\frac{\alpha_1^g}{\left(\sum_{g'} W_j^{g'} \right)} + \frac{\alpha_2^{gg}}{W_j^g} \right] \tag{A4.17}$$

We can then substitute this in Equation (A4.14):

$$\frac{\partial S_{ij}^g}{\partial W_j^g} = S_{ij}^g \left(1 - \frac{S_{ij}^g}{e_i^g P_i} \right) \left[\frac{\alpha_1^g}{\left(\sum_{g'} W_j^{g'} \right)} + \frac{\alpha_2^{gg}}{W_j^g} \right] \tag{A4.18}$$

Appendix 4

The crucial term is that in square brackets. The denominator
of the first term is a sum of W_j^g's and will not vanish unless
some location has no faciltiies at all. (We should also
emphasise that the expression $\sum_{g'} W_j^{g'}$ occurs in S_{ij}^g raised to the
power α_1^g, and so there will not be any problem at all unless
$\alpha_1^g > 1$.) In this case, W_j^1 would also vanish. We can, therefore, mainly concentrate on the second term: this appears as a
factor in S_{ij}^g, through W_j^g, raised to the power α_2^{gg}. Hence, the
usual kinds of results apply: if $\alpha_2^{gg} > 1$, then this term tends
to infinity as $W_j^g \to 0$, and we saw earlier that this means that
there will be no stable zero equilibrium point, and no jumps
when parameters change. In effect, what this analysis means is
that, once a lower order centre exists at a location, then jumps
are only possible in relation to higher order centres than that.
This is a useful result for an evolutionary model.

REFERENCES AND BIBLIOGRAPHY

Agergard, E.,Olson, P.A. and Allpass, J. (1970) The interaction between retailing and the urban centre structure: a
theory of spiral movement, *Environment and Planning, 2, 55-71.*

Alexander, D. (1979) Catastrophic misconception?, *Area, 11,*
228-9.

Allen, P.M. (1976) Evolution, population dynamics and
stability, *Proceedings, National Academy of Sciences of the
USA, 73,* 665-8.

Allen, P.M. and Sanglier, M. (1979) A dynamic model of growth
in a central place system, *Geographical Analysis, 11,* 256-72.

Allen, P.M., Deneubourg, J.L., Sanglier, M., Boon, F. and
de Palma, A. (1978) The dynamics of urban evolution, Vol. 1:
Interurban evolution; Vol. 2: Intraurban evolution, *Final
Report to the US Department of Transportation,* Washington DC.

Amson, J.C. (1972-A) The dependence of population distribution
on location costs, *Environment and Planning, 4,* 163-81.

Amson, J.C. (1972-B) Equilibrium models of cities: 1. An
axiomatic theory, *Environment and Planning, 4,* 429-44.

Amson, J.C. (1973) Equilibrium models of cities: 2. Single
species cities, *Environment and Planning, 5,* 295-338.

Amson, J.C. (1974) Equilibrium and catastrophic modes of urban
growth, in E.L. Cripps (Ed.) *Space-time concepts in urban
and regional models,* Pion, London, 108-28.

Amson, J.C. (1975) Catastrophe theory: a contribution to the
study of urban problems, *Environment and Planning, B, 2,*
177-221.

Amson, J.C. (1977) A note on civic state equations, *Environment
and Planning A, 9,* 105-10.

Amson, J.C. (1979) Private communication.

Angel, S. and Hyman, G.M. (1976) *Urban fields,* Pion, London.

Atkins, R. (1977) *Combinatorial connectivities in social
systems,* Birkhauser, Barle.

Atkinson, G. (1976) Catastrophe theory in geography - a new
look at some old problems, Mimeo, Department of Geography,
University of Cambridge.

Ayala, F.J. (1978) The mechanisms of evolution, *Scientific
American, 239*(3), 56-69.

References

Baker, A.R.H. (1978) Settlement pattern evolution and catastrophe theory: a comment, *Transactions, Institute of British Geographers*, New Series, *4*, 435-44.

Balasko, Y. (1975) Some results on uniqueness and stability of equilibrium in general equilibrium theory, *Journal of Mathematical Economics*, *2*, 95-118.

Balasko, Y. (1978) The behaviour of economic equilibria: a catastrophe theory approach, *Behavioural Sciences*, *23*, 375-82.

Batty, M. (1976) *Urban modelling*, Cambridge University Press, Cambridge.

Baumol, W.J. (1959) *Economic dynamics*, 2nd edn, Macmillan, New York.

Beaumont, J.R. and Clarke, M. (1979) Improving supply side representations in urban models, with specific reference to central place and Lowry models, *Sistemi Urbani*, forthcoming.

Beavon, K.S.O. (1977) *Central place theory: a reinterpretation*, Longmans, London.

Beddington, J.R., Free, C.A. and Lawton, J.H. (1975) Dynamic complexity in predator-prey models formed in difference equations, *Nature*, *255*, 58-60.

Bellman, R. (1968) *Some vistas of modern mathematics*, University of Kentucky Press, Lexington, Kentucky.

Berry, B.J.L. (1967) *Geography of market centres and retail distribution*, Prentice-Hall, Englewood Cliffs, New Jersey.

Berry, B.J.L. and Garrison, W.L. (1958) Recent developments of central place theory, *Papers, Regional Science Association*, *4*, 107-20.

Blase, J.H. (1979) Hysteresis and catastrophe theory: empirical identification in transport modelling, *Environment and Planning A*, *11*, 675-88.

Boddington, T., Gray, P. and Wake, G.C. (1977) Criteria for thermal explosions with and without reactant consumption, *Proceedings of the Royal Society of London A*, *357*, 403-22.

Boon, F. and de Palma, A. (1978) Boolean formalism and urban development, Mimeo, Service de Chimie-Physique II, Universite Libre de Bruxelles.

von Boventer, E. (1976) Transportation costs, accessibility and agglomeration economies: centres, subcentres and metropolitan structures, *Papers, Regional Science Association*, *37*, 169-83.

Bowers, R.G. (1978-A) Statistical dynamic models of social systems I: the general theory, *Behavioural Science*, *23*, 109-19.

Bowers, R.G. (1978-B) Statistical dynamic models of social systems II: discontinuity and conflict, *Behavioural Science*, *23*, 120-9.

Broadbent, T.A. (1977) *Planning and profit in the urban economy*, Methuen, London.

Brunet, R. (1970) *Les phénomènes de discontinuité en géographie*, Editions du Centre National de la Recherche Scientifique, Paris.

Casetti, E. (1970) Equilibrium population partitions between urban and agricultural occupations, Mimeo, Department of Geography, Ohio State University.

Casti, J. and Swain, H. (1975) Catastrophe theory and urban processes, RM-75-14, IIASA, Laxenburg, Austria.

Chillingworth, D.R.J. (1975) Elementary catastrophe theory, *Bulletin of the Institute of Mathematics and its Applications*, *11*, 155-9.

Chillingworth, D.R.J. (1976) *Differential topology with a view to applications*, Pitman, London.

Christaller, W. (1933) *Die zentralen orte in Suddendeutschland*, Jena; English translation by C.W. Baskin, *Central places in southern Germany*, Prentice-Hall, Englewood Cliffs.

Chudzynska, I. and Slodkowski, Z. (1979) Comments on the urban spatial-interaction model based on the intervening opportunities principle, *Environment and Planning A*, *11*, 527-39.

Clark, C.W. (1976) *Mathematical bioeconomics*, John Wiley, New York.

Clarke, M., Keys, P. and Williams, H.C.W.L. (1979) Household dynamics and economic forecasting: a micro-simulation approach, Working Paper 257, School of Geography, University of Leeds.

Coelho, J.D. (1977) The use of mathematical optimisation methods in model-based land-use planning. An application to the new town of Santo Andre, PhD thesis, School of Geography, University of Leeds.

Coelho, J.D. and Williams, H.C.W.L. (1977) On the design of land use plans through locational surplus maximisation, *Papers, Regional Science Association*, *40*, 71-85.

Coelho, J.D., Williams, H.C.W.L. and Wilson, A.G. (1978) Entropy maximising submodels within overall mathematical programming frameworks: a correction, *Geographical Analysis*, *10*, 195-201.

Coelho, J.D. and Wilson, A.G. (1976) The optimum size and location of shopping centres, *Regional Studies*, *10*, 413-21.

Coelho, J.D. and Wilson, A.G. (1977) An equivalence theorem to integrate entropy maximising models within overall mathematical programming frameworks, *Geographical Analysis*, *9*, 160-73.

References

Cooke, J. and Zeeman, E.C. (1976) A clock and wavefront model for control of the number of repeated structures during animal morphogenesis, *Journal of Theoretical Biology, 58,* 455-76. Reprinted in E.C. Zeeman (1977) op. cit, 235-56.

Cooke, K. and Renfrew, C. (eds.) (1978) *Transformations: mathematical approaches to culture change,* Academic Press, London.

Cordey-Hayes, M. (1972) Dynamic framework for spatial models, *Socio-Economic Planning Sciences, 6,* 365-85.

Cordey-Hayes, M. (1975) Migration and the dynamics of multi-regional population systems, *Environment and Planning A, 7,* 793-814.

Cowan, J.D. (1970) A statistical mechanics of nervous activity, in *Some mathematical questions in biology,* American Mathematical Society, Providence, Rhode Island, 1-57.

Dacey, M. (1976) *An introduction to the mathematical theory of central places,* 1. Northwestern University Press, Evanston.

Day, M. and Tivers, J. (1979) Catastrophe theory and geography: a Marxist critique, *Area, 11,* 54-8.

Day, M. and Tivers, J. (1979) Catastrophic misconceptions?, *Area, 11,* 231.

Dendrinos, D.S. (1977-A) Short-run disequilibria in urban spatial structures, *Regional Science Perspectives, 7*(2), 37-41.

Dendrinos, D.S. (1977-B) Slums in capitalist urban settings: some insights from catastrophe theory, mimeo; published, 1979, *Geographia Polonica, 42,* 63-75.

Dendrinos, D.S. (1978-A) Operating speeds and volume to capacity ratios: the observed relationship and the fold catastrophe, *Transportation Research, 12,* 191-4.

Dendrinos, D.S. (1978-B) Urban dynamics and urban cycles, *Environment and Planning A, 10,* 43-9.

Dendrinos, D.S. and Mullally, H. (1979) Fast and slow equations: the development pattern of urban settings, presented to the Annual Meeting of the North American Section, Regional Science Association, Los Angeles.

Deneubourg, J.L., de Palma, A. and Kahn, D. (1979) Dynamics models of competition between transport modes, *Environment and Planning A, 11,* 665-73.

Eilon, S., Tilley, R.P.R. and Fowkes, T.R. (1969) Analysis of a gravity demand model, *Regional Studies, 3,* 115-22.

Evans, S.P. (1973) A relationship between the gravity model for trip distribution and the transportation problem in linear programming, *Transportation Research, 7,* 39-61.

Ferguson, J.A. (1976) Investment decisions and sudden changes in transport, *Surveyor,* 9 July, 10-11.

References

Field, R.J. (1975) Limit cycle oscillations in the reversible oregonator, *Journal of Chemical Physics*, *63*, 2289-96.

Field, R.J. and Noyes, R.M. (1974) Oscillations in chemical systems IV. Limit cycle behaviour in a model of a real chemical reaction, *Journal of Chemical Physics*, *60*, 1877-944.

Forrester, J.W. (1968) *Principles of systems*, Wright-Allen Press, Cambridge, Mass.

Fowler, D.H. (1972) The Riemann-Hugoniot catastrophe and van der Waals equations, in Waddington, C.H. (Ed.) *Towards a theoretical biology*, Edinburgh University Press, Edinburgh, 1-7.

Gardner, M.R. and Ashby, W.R. (1970) Connectance of large dynamical (cybernetic) systems: critical values for stability, *Nature*, *228*, 784.

Ginsberg, R.B. (1973) Stochastic models of residential and geographic mobility for heterogeneous populations, *Environment and Planning*, *5*, 113-24.

Ginsberg, R.B. (1978) Probability models of residence histories: analysis of time between moves, In Clark, W.A.V. and Moore, E.G. (Eds.) op cit., 233-65.

Glass, L. and Kauffman, S.A. (1973) The logical analysis of continuous non-linear biochemical control networks, *Journal of Theoretical Biology*, *39*, 103-29.

Gleave, D. and Cordey-Hayes, M. (1977) *Migration dynamics and labour market turnover*, Pergamon, Oxford.

Goering, J.M. (1978) Neighbourhood tipping and racial transition: a review of social science evidence, *Journal of the American Institute of Planners*, *44*, 68-78.

Goodwin, P.B. (1977) Habit and hysteresis in modal choice, *Urban Studies*, *14*, 95-8.

Goodwin, R.M. (1951) The non-linear accelerator and the persistance of business cycles, *Econometrics*, *19*, 1-17.

Gray, P., Griffiths, J.F. and Moule, R.J. (1974) Thermokinetic oscillations accompanying propane oxidation, *Farraday Symposium*, *9*, Chemical Society, London.

Hagerstrand, T. (1967) *Innovation diffusion as a spatial process*, University of Chicago Press, Chicago, Ill.

Hansen, W.G. (1959) How accessibility shapes land use, *Journal of the American Institute of Planners*, *25*, 73-6.

Harris, B. (1965) A model of locational equilibrium for the retail trade, Mimeo, Institute for Urban Studies, University of Pennsylvania.

Harris, B. and Wilson, A.G. (1978) Equilibrium values and dynamics of attractiveness terms in production-constrained spatial-interaction models, *Environment and Planning A*, *10*, 371-88.

References

Hassell, M.D. (1978) *The dynamics of arthropod predator-prey systems*, Princeton University Press, Princeton, New Jersey.

Hirsch, M.W. and Smale, S. (1974) *Differential equations, dynamical systems and linear algebra*, Academic Press, New York.

Hoppensteadt, F.C. (1978) Mathematical aspects of population biology, in L.A. Stean (Ed.) *Mathematics today: twelve informal essays*, Springer Verlag, New York, 297-320.

Hotelling, H. (1929) Stability in competition, *Economic Journal, 39,* 41-57.

Huff, D.L. (1964) Defining and estimating a trading area, *Journal of Marketing, 28,* 34-8.

Innis, G. (1974) Dynamic analysis in 'soft science' studies: in defence of difference equations, in Levin, S. (Ed.) *Mathematical problems in biology – Victoria Conference*, Springer Verlag, New York, 102-22.

Isard, W. (1976) On hierarchical dynamics, mimeo, Department of Regional Science, University of Pennsylvania.

Isard, W. (1977) Strategic elements of a theory of major structure change, *Papers, Regional Science Association, 38,* 1-14.

Isard, W. and Liossatos, P. (1977) Models of transition processes, *Papers, Regional Science Association, 39,* 27-59.

Isard, W. and Liossatos, P. (1978) A simplistic multiple growth model, *Papers, Regional Science Association, 41,* 7-13.

Isard, W. and Liossatos, P. (1979) *Spatial dynamics and optimal space-time developments*, North Holland, Amsterdam.

Jordan, D.W. and Smith, P. (1977) *Non-linear ordinary differential equations*, Oxford University Press, Oxford.

Kauffman, S.A. (1969) Metabolic stability and epigenesis in randomly constructed genetic sets, *Journal of Theoretical Biology, 22,* 347-67.

Kauffman, S.A. (1977) Chemical patterns, compartments and a binary epigenetic code in Crosophila, *American Zoologist, 17,* 631-48.

Kerner, E.H. (1972) *Gibbs ensemble: biological ensemble*, Gordon and Breach, New York.

Keys, P. (1977) A relationship between digraphs and differential equations with an application to an urban system, Working Paper 194, School of Geography, University of Leeds.

Kilmister, C.W. (1976) Population in cities, *The Mathematical Gazette, 60,* 11-24.

Ladde, G.S. (1977) Competitive processes II. Stability of random systems, *Journal of Theoretical Biology, 68,* 331-54.

Lakshmanan, T.R. and Hansen, W.G. (1965) A retail market potential model, *Journal of the American Institute of Planners*, *31*, 134-43.

Leon, J.A. (1974) Selection in contexts of interspecific competition, *The American Naturalist*, *108*, 739-55.

Leonardi, G. (1973) Localizzazione ottimale dei servizi urbani, *Ricerca Operativa*, *12*, 15-43.

Leonardi, G. (1975) Un nuove algoritmo per il problema bella localizzazione ottimale dei servizi urbani, *Atti delle Giornate, Associazione Italiana di Recerca Operativa*, Milano, 121-32.

Leonardi, G. (1977) Analogie meccanico-statisticle nei modelli di interazione spaziole, *Atti delle Giornate, Associazione Italiana di Recerca Operativa*, Parma, 530-9.

Leonardi, G. (1978) Optimum facility location by accessibility maximising, *Environment and Planning A*, *10*, 1287-305.

Levin, S. (1970) Community equilibria and stability, and an extension of the competitive exclusion principle, *The American Naturalist*, *104*, 413-23.

Levins, B.R. (1971) The operation of selection in situations of interspecific competition, *Evolution*, *25*, 249-64.

Levins, R. (1968) *Evolution in changing environments*, Princeton University Press, New Jersey.

Lösch, A. (1943) *Die raumliche ordnung der wirtschaft*, Fischer, Jena; translated by W.G. Woglom (1954) *The economics of location*, Yale University Press, New Haven.

Lowry, I.S. (1964) *A model of metropolis*, RM-4035-RC, Rand Corporation, Santa Monica.

McFadden, D. (1974) Conditional logit analysis of qualitative choice behaviour, in Zarembka, P. (Ed.) *Frontiers of econometrics*, Academic Press, New York.

Macgill, S.M. (1977) The Lowry model as an input-output model and its extensions to incorporate full inter-sectoral relations, *Regional Studies*, *11*, 337-54.

Macgill, S.M. and Wilson, A.G. (1979) Equivalences and similarities between some alternative urban and regional models, *Sistemi Urbani*, *1*, 9-40.

May, R.M. (1971) Stability in multi-species community models, *Mathematical Biosciences*, *12*, 59-79.

May, R.M. (1973) *Stability and complexity in model ecosystems*, Princeton University Press, Princeton, New Jersey.

May, R.M. (1974) Biological populations with nonoverlapping generations: stable points, stable cycles and chaos, *Science*, *196*, 645-7.

References

May, R.M. (1975) Biological populations obeying difference equations: stable points, stable cycles and chaos, *Journal of Theoretical Biology, 51,* 511-24.

May, R.M. (1976-A) Simple mathematical models with very complicated dynamics, *Nature, 261,* 459-67.

May, R.M. (Ed.) (1976-B) *Theoretical ecology,* Blackwell, Oxford.

May, R.M. (1978) The evolution of ecological systems, *Scientific American, 239*(3), 161-75.

May, R.M. and Oster, G.F. (1976) Bifurcation and dynamic complexity in simple ecological models, *The American Naturalist, 110,* 573-99.

Maynard Smith, J. (1974) *Models in ecology,* Cambridge University Press, Cambridge.

Maynard Smith, J. (1978) The evolution of behaviour, *Scientific American, 239*(3), 176-92.

Mayr, E. (1978) Evolution, *Scientific American, 239*(3), 46-55.

Mees, A.I. (1975) The revival of cities in medieval Europe - an application of catastrophe theory, *Regional Science and Urban Economics, 5,* 403-26.

Nicolis, G. and Prigogine, I. (1977) *Self-organisation in non-equilibrium systems,* John Wiley, New York.

O'Flaherty, C.A. (1974) *Highways, Vol. 1: Highways and traffic,* Edward Arnold, London.

Orcutt, G.H., Greenberger, M., Korbel, J. and Rivelen, A.M. (1961) *Microanalysis of socio-economic systems: a simulation study,* Harper and Row, New York.

Oster, G. and Guckenheimer, J. (1976) Bifurcation phenomena in population models, in Marsden, J.E. and McCracken, M. (Eds.) *The Hopf bifurcation and its applications,* Springer Verlag, New York.

Oster, G. and Takahashi, Y. (1974) Models for age-specific interactions in a periodic environment, *Ecological Monographs, 44,* 483-501.

de Palma, A., Stengers, I. and Pahout, S. (1979) Boolean equations with temporal delays, Mimeo, Service de Chimie-Physique II, Universite Libre de Bruxelles.

Papageorgiou, G.J. (1979) Private communication.

Papageorgiou, G.J. (1980) On sudden urban growth, *Environment and planning A, 12,* 1035-50.

Phiri, P.A. (1979) Equilibrium points and control problems in dynamics urban modelling, PhD thesis, University of Leeds.

Phiri, P.A. (1980) Calculation of the equilibrium configuration of shopping facilty size, *Environment and Planning A, 12,* 983-1000.

Pielou, E.C. (1974) Competition on an environmental gradient, in Levin, S. (Ed.) *Mathematical problems in biology - Victoria Conference,* Springer, Verlag, New York, 184-204.

Pitt, D.H. and Poston, T. (1979) Determinacy and unfoldings in the presence of a boundary, mimeo.

Poston, T. and Stewart, I.N. (1978) *Catastrophe theory and its applications,* Pitman, London.

Poston, T. and Wilson, A.G. (1977) Facility size vs. distance travelled: urban services and the fold catastrophe, *Environment and Planning A, 9,* 681-6.

Prigogine, I., Nicolis, G. and Babloyantz, A. (1972) Thermodynamics of evolution, *Physics Today,* November, 23-8.

Putman, S. and Ducca, F. (1978) Private communication.

Rees, P.H. and Wilson, A.G. (1977) *Spatial population analysis,* Edward Arnold, London; Academic Press, New York.

Renfrew, A.C. and Poston, T. (1978) Discontinuous change in settlement pattern - a processual analysis, in Renfrew, A.C. and Cooke, K.L. (Eds.) *Transformations: a mathematical approach to culture change,* Academic Press, London.

Rescigno, A. and Richardson, I. (1967) Struggle for life I: two species, *Bulletin of Mathematical Biophysics, 29,* 377-88.

Richardson, H.W. (1975-A) Two disequilibrium models of regional growth, in Cripps, E.L. (Ed.) *Space-time concepts in urban and regional models,* Pion, London, 46-55.

Richardson, H.W. (1975-B) Discontinuous densities, urban spatial structure and growth: a new approach, *Land Economics, 51,* 305-15.

Richardson, H.W. (1977) *The new urban economics: and alternatives,* Pion, London.

Richardson, L.F. (1960) *Arms and insecurity,* The Boxwood Press, Pittsburgh.

Roberts, F.S. (1976) *Discrete mathematical models,* Prentice Hall, Englewood Cliffs, New Jersey.

Samuelson, P.A. (1947) *Foundations of economic analysis,* Harvard University Press, Cambridge, Mass.

Senior, M.L. (1973) Approaches to residential location modelling 1: urban ecological and spatial interaction models, *Environment and Planning, 5,* 165-97.

References

Senior, M.L. (1974) Approaches to residential location modelling 2: urban economic models and some recent developments, *Environment and Planning A, 6,* 369-409.

Senior, M.L. and Wilson, A.G. (1974) Some explorations and syntheses of linear programming and spatial interaction models of residential location, *Geographical Analysis, 6,* 209-37.

Smith, T.R. (1976) The 'internal referencing' structural instability in an environment characterised by random variation: the example of the WYC banks in the crisis of 1857, Mimeo, Department of Geography, University of Santa Barbara.

Smith, T.R. (1977) Continuous and discontinuous response to smoothly decreasing effective distance: an analysis with special reference to 'overbanking' in the 1920s, *Environment and Planning A, 9,* 461-75.

Sprinkhuizen-Kuyper, I.G. (1976) Bifurcation of periodic solutions and a mathematical model for the struggle between antigen and antibody, in Temme, N.M. (Ed.) *Non-linear analysis, 1,* Mathematische Centrum, Amsterdam.

Steele, J. (1974) Stability of plankton ecosystems, in Usher, M.B. and Williamson, M.H. (Eds.) *Ecological stability,* Chapman and Hall, London, 179-91.

Stewart, I.N. (1979) Letter to the Editor, *Nature, 270,* 382.

Thom, R. (1975) *Structural stability and morphogenesis,* W.A. Benjamin, Reading, Mass.

Thomas, R. (1973) Boolean formalisation of genetic control circuits, *Journal of Theoretical Biology, 42,* 563-85.

Tomlin, S.G. (1969) A kinetic theory of traffic distribution and similar problems, *Environment and Planning, 1,* 221-2.

Tomlin, S.G. (1979) A kinetic theory of urban dynamics, *Environment and Planning A, 11,* 97-106.

Turing, A.M. (1952) The chemical basis of morphogenesis, *Proceedings of the Royal Society B, 237,* 37-72.

Turner, J.E. and Rapport, D.J. (1974) An economic model of population growth and competition in natural communities, in Levin, S. (Ed.) *Mathematical problems in biology - Victoria Conference,* Springer Verlag, New York, 236-40.

Usher, M.B. and Williamson, M.H. (Eds.) (1974) *Ecological stability,* Chapman and Hall, London.

Varaprasad, N. (1979) Transport costs and the dynamics of population redistribution: two strategic models for the South-East, PhD thesis, Centre for Transport Studies, Cranfield Institute of Technology, Bedford.

Varian, H.L. (1979) Catastrophe theory and the business cycle, *Economic Inquiry, 17,* 14-28.

Wagstaff, J.M. (1976) Some thoughts about geography and catastrophe theory, *Area, 8,* 316-20.

Wagstaff, J.M. (1978) A possible interpretation of settlement pattern evolution in terms of catastrophe theory, *Transactions, Institute of British Geographers,* New Series, *3,* 165-78.

Wagstaff, J.M. (1979-A) Catastrophic misconception?, *Area, 11,* 230-1.

Wagstaff, J.M. (1979-B) Dialectical materialism, geography and catastrophe theory, *Area, 11,* 326-33.

Weaver, W. (1958) A quarter century in the natural sciences, *Annual Report,* The Rockefeller Foundation, New York, 7-122.

Webber, M.J. (1980) *Information theory and urban spatial structure,* Croom Helm, London.

Webber, M.J. and Joseph, A.E. (1978) Spatial diffusion processes 1: a model and an approximation method, *Environment and Planning A, 10,* 651-65.

Webber, M.J. and Joseph, A.E. (1979) Spatial diffusion processes 2: numerical analysis, *Environment and Planning A, 11,* 335-47.

Wheaton, W.C. (1974) A comparative static analysis of urban spatial structure, *Journal of Economic Theory, 9,* 223-37.

White, R.W. (1974) Sketches of a dynamic central place theory, *Economic Geography, 50,* 219-27.

White, R.W. (1977) Dynamic central place theory: results of a simulation approach, *Geographical Analysis, 9,* 226-43.

White, R.W. (1978) The simulation of central place dynamics: two sector systems and the rank-size distribution, *Geographical Analysis, 10,* 201-8.

Williams, H.C.W.L. (1977) On the formation of travel demand models and economic evaluation measures of user benefit, *Environment and Planning A, 9,* 285-344.

Wilson, A.G. (1968) Models in urban studies: a synoptic review of recent literature, *Urban Studies, 5,* 249-76.

Wilson, A.G. (1970) *Entropy in urban and regional modelling,* Pion, London.

Wilson, A.G. (1974) *Urban and regional models in geography and planning,* John Wiley, Chichester.

Wilson, A.G. (1976-A) Catastrophe theory and urban modelling: an application to modal choice, *Environment and Planning A, 8,* 351-6.

References

Wilson, A.G. (1976-B) Retailers' profits and consumers' welfare in a spatial interaction shopping model, in Masser, I. (Ed.) *Theory and practice in regional science*, Pion, London, 42-59.

Wilson, A.G. (1977-A) Spatial interaction and settlement structure: towards an explicit central place theory, in A. Karlqvist, L. Lundqvist, F. Snickars and J. Weibull (Eds.) *Spatial interaction theory and planning models*, North Holland, Amsterdam, 137-56.

Wilson, A.G. (1977-B) Recent developments in urban and regional modelling: towards an articulation of systems' theoretical foundation, *Giornate di Lavoro*, AIRO, Parma.

Wilson, A.G. (1978) Towards models of the evolution and genesis of urban structure, in Bennett, R.J., Martin, R.C. and Thrift, N.J. (Eds.) *Dynamic analysis of spatial systems in geography and regional science*, Pion, London, 79-90.

Wilson, A.G. (1979-A) Aspects of catastrophe theory and bifurcation theory in regional science, Working Paper 249, School of Geography, University of Leeds.

Wilson, A.G. (1979-B) Equilibrium and transport systems dynamics, in Hensher, D. and Stopher, P. (Eds.) *Behavioural travel modelling*, Croom Helm, London, 164-86.

Wilson, A.G. (1979-C) Residential mobility policy, models and information, Working Paper 251, School of Geography, University of Leeds.

Wilson, A.G. and Clarke, M. (1979) Some illustrations of catastrophe theory applied to urban retailing structures, in Breheny, M. (Ed.) *London Papers in Regional Science*, Pion, London, 5-27.

Wilson, A.G. and Kirkby, M.J. (1980) *Mathematics for geographers and planners*, 2nd edn, Oxford University Press, Oxford, Chapter 10.

Wilson, A.G. and Macgill, S.M. (1979) A systems analytical framework for comprehensive urban and regional modelling, *Geographica Polonica*, *42*, 9-25.

Wilson, A.G. and Pownall, C.E. (1976) A new representation of the urban system for modelling and for the study of micro-level interdependence, *Area*, *8*, 246-54.

Wilson, A.G. and Senior, M.L. (1974) Some relationships between entropy maximising models, mathematical programming models and their duals, *Journal of Regional Science*, *14*, 207-15.

Woods, R.I. (1977) Population turnover, tipping points and Markov chains, *Transactions, Institute of British Geographers, New Series*, *2*, 473-89.

Yamane, T. (1968) *Mathematics for economists*, Prentice-Hall, Englewood Cliffs, New Jersey.

Zahler, R.S. and Sussman, H.J. (1977) Claims and accomplishments of applied catastrophe theory, *Nature, 269,* 759-63.

Zeeman, E.C. (1974) Primary and secondary waves in developmental biology, in *Lectures on Mathematics in the Life Sciences, 7,* American Mathematical Society, Providence, Rhode Island, 69-161. Reprinted in Zeeman, E.C. (1977) op cit, 141-233.

Zeeman, E.C. (1977-A) *Catastrophe theory,* Addison-Wesley, Reading, Mass.

Zeeman, E.C. (1977-B) Letter to the editor, *Nature, 270,* 381.

Zovick, M. (1978) Dialectics and catastrophe, in Geyer, R.F. and van der Zouven, J. (Eds.) *Sociocybernetics, 1,* Martinus Nijhoff, Leiden, 129-54.

AUTHORS INDEX

Allen, P.M., 113, 116, 157 *et seq.*, 170, 173, 176, 235, 264
Amson, J.C., 14, 16, 18, 28, 67, 68, 69 *et seq.*, 92, 140
Angel, S., 202
Ashby, W.R., 237
Ayala, F.J., 234

Babloyantz, A., 235
Balasko, Y., 248
Batty, M., 180
Baumol, W.J., 248
Beaumont, J.R., 190
Beavon, K.S.O., 174, 176
Beddington, J.R., 245, 265
Bellman, R., 38, 62
Berry, B.J.L., 174, 176
Blase, J.H., 8, 210
Boon, F., 256 *et seq.*
von Böventer, E., 175
Broadbent, T.A., 189

Casetti, E., 89
Casti, J., 12, 21, 67, 74 *et seq.*, 81, 82, 88, 92
Chillingworth, D.R.J., 14
Christaller, W., 173, 177
Clark, C.W., 249 *et seq.*
Clarke, M., 135 *et seq.*, 190, 223
Coelho, J.D., 280, 283 *et seq.*, 285
Cooke, J., 234, 247
Cordey Hayes, M., 169
Cowan, J.D., 237

Dendrinos, D.S., 67, 88 *et seq.*, 92, 219 *et seq.*
Deneubourg, J.L., 210 *et seq.*
Ducca, F., 114

Evans, S.P., 192, 278

Field, R.J., 230
Forrester, J.W., 245
Fowler, D.H., 70
Free, C.A., 245, 265

Gardner, M.R., 237
Garrison, W.L., 176
Ginsberg, R.B., 169
Glass, L., 233 *et seq.*, 256

Index

Gleave, D., 169
Goodwin, P.B., 8, 207 *et seq.*
Goodwin, R.M., 253

Hägerstrand, T., 268
Hansen, W.G., 96, 282
Harris, B., 10, 96, 117, 118, 141, 262, 264
Hassell, M.D., 245
Hirsch, M.W., 43, 46, 51, 151 *et seq.*
Hoppensteadt, F.C., 253
Hotelling, H., 198
Huff, D.L., 96
Hyman, G.M., 202

Innis, G., 245
Isard, W., 67, 82 *et seq.*, 92, 148

Jordan, D.W., 33, 38
Joseph, A.E., 268

Kahn, D., 210 *et seq.*
Kauffman, S.A., 233 *et seq.*, 256
Kerner, E.H., 248
Keys, P., 213
Kirkby, M.J., 206

Lakshmanan, T.R., 96
Lawton, J.H., 245, 265
Leon, J.A., 235
Leonardi, G., 282
Levin, S., 238
Levins, B.R., 235
Levins, R., 235
Liossatos, P., 148
Lösch, A., 177
Lowry, I.S., 94, 99, 103, 108, 116, 148, 179, 180, 185, 189, 285

Macgill, S.M., 2, 189, 285
May, R.M., 235, 237 *et seq.*, 240 *et seq.*, 253, 260 *et seq.*
Maynard Smith, J., 43, 46, 49, 235
Mayr, E., 234
Mees, A.I., 36, 67, 82 *et seq.*, 88, 92

Nicolis, G., 156, 226 *et seq.*, 232, 235
Noyes, R.M., 230

O'Flaherty, C.A., 219
Oster, G., 239, 240

Pahout, S., 256 *et seq.*
de Palma, A., 210 *et seq.*, 256 *et seq.*
Papageorgiou, G.J., 67, 89 *et seq.*, 92
Phiri, P.A., 118, 172

316

Pielou, E.C., 238, 246, 247
Pitt, D.H., 31
Poston, T., 7, 14, 28, 30, 31, 67, 78 *et seq.*, 84, 95, 142, 220
    232, 233, 247
Prigogine, I., 62, 156 *et seq.*, 226 *et seq.*, 232, 235
Putman, S., 114

Rees, P.H., 240
Richardson, H.W., 190
Richardson, L.F., 248
Roberts, F.S., 238

Samuelson, P.A., 248
Sanglier, M., 157 *et seq.*, 264
Senior, M.L., 104, 192, 278
Smale, S., 43, 46, 51, 151
Smith, P., 33, 38
Smith, T.R., 248
Sprinkhuizen-Kuyper, I.G., 234
Steele, J., 238
Stengers, I., 256 *et seq.*
Stewart, I.N., 7, 14, 28, 30, 31, 142, 232, 233, 247
Sussman, H.J., 92
Swain, H., 12, 21, 67, 74 *et seq.*, 81, 82, 88, 92

Takahashi, Y., 239
Thom, R., 1, 3, 14, 15, 17, 21, 22, 30, 140, 247
Thomas, R., 256
Tomlin, S.G., 169
Turing, A.M., 234

Varaprasad, N., 169
Varian, H.L., 253

Wagstaff, J.M., 67, 84 *et seq.*, 92
Weaver, W., 225
Webber, M.J., 268, 271
Wheaton, W.C., 89
White, R.W., 114, 141, 263
Williams, H.C.W.L., 223, 280, 283 *et seq.*, 285 *et seq.*
Wilson, A.G., 2, 6, 10, 60, 67, 78 *et seq.*, 84, 92, 95, 96, 97,
    104, 109, 116, 118, 135 *et seq.*, 142, 169, 178, 192, 198,
    199, 202, 206 *et seq.*, 220, 240, 262, 264, 271 *et seq.*, 278,
    280 *et seq.*, 285

Yamane, T., 131

Zahler, R.S., 92
Zeeman, E.C., 7, 24, 28, 57, 58, 60, 198, 204, 222, 232, 234,
    246, 247, 267

SUBJECT INDEX

Accessibility maximisation, 108, 282
Accessibility to shops, 106
Age-disaggregation in geographic models, 240
Antigen-antibody reactions, 234
Attractiveness factors, 165, 292
    composite, 105 *et seq.*, 191, 297
    in disaggregated retail model, 102
    relative, 160
    residential, 106
Attractors, 33, 39
Autocatalytic reactions, 226

Balancing mechanisms, 10, 97
    in disaggregated retail model, 149
Banks, 98, 248
Behaviour manifold, 20, 35
Behaviour of system, 4
Belousov-Zhabotinsky reaction, 230 *et seq.*, 233
Bid rent functions, 190
Bifurcation, 2, 42
    and differential equations, 33 *et seq.*
    and modal choice, 210 *et seq.*
    at the meso scale, 93 *et seq.*, 155 *et seq.*
    properties, 62
    set, 20
Bimodal distributions, 204
Biochemistry, 233 *et seq.*
Biology, 232 *et seq.*
    evolutionary, 234 *et seq.*
Boolean models, 233 *et seq.*, 256 *et seq.*
    urban, state table for, 259
Branching points, 36
Brussels School, 156 *et seq.*, 228 *et seq.*
Business cycles, 253 *et seq.*
Butterfly catastrophe
    and property prices, 76
    and revival of cities, 82 *et seq.*
    and trimodal distributions, 204
    control manifold for, 77

Canonical form
    as a model, 18
    of surfaces, 5
Canonical surface, 92
Capacity functions, 199
Catastrophe, butterfly
    and property prices, 76
    and revival of cities, 82
    control manifold for, 77

Index

Catastrophe, cusp, 25 *et seq.*
 and Amson's fourth law, 72
 and bimodal distributions, 204
 and business cycles, 253 *et seq.*
 and central place theory, 75
 and city growth, 12
 and depressions, 255
 and genetic switching, 233
 and modal choice, 8 *et seq.*, 206 *et seq.*
 and property prices, 76
 and settlement evolution, 86
 and structural change, 82 *et seq.*
 and sudden urban growth, 89 *et seq.*
 and travelling frontier, 247
 in chemistry, 227
Catastrophe, fold, 22 *et seq.*
 and Amson's third law, 71
 and central place theory, 75
 and centre size, 80
 and fisheries yields, 252
 and Lotka-Volterra equations, 48
 and retail centre size, 129
 and spatial structure, 10
 and speed-flow relationships, 218 *et seq.*
Catastrophe machine, Zeeman, 7, 27
Catastrophe, mushroom, 88
Catastrophe, parabolic umbilic, 88
Catastrophe, Riemann-Hugoniot, 70
Catastrophe set, 20
Catastrophe theory
 and differential equations, 58 *et seq.*
 and modal choice, 206 *et seq.*
 in ecology, 246
 mathematics of, 1 *et seq.*
 possibilities of, 4
Catastrophes, 5, 20 *et seq.*
 constraint, 30
 dual, 28
 elementary, table of, 29
 higher order, 28 *et seq.*
Catastrophic modes of urban growth, 69 *et seq.*
Cell biochemistry, 233 *et seq.*
Central places, functional levels of, 13, 75
Central place theory, 12, 74 *et seq.*, 196
 new, 173 *et seq.*
Centre size and location, 78 *et seq.*, 95 *et seq.*
 and control theory, 172
 and the fold catastrophe, 129
 benefits of, 106
Centres, 175
 evolution of, 198 *et seq.*
 types of, 177
Centres, shopping
 and difference equations, 260 *et seq.*

Change
  mechanisms of, 61
  relative speeds of, 59 *et seq.*
Chaotic behaviour, 41, 264
Chemical concentrations, 227
Chemistry, physical, 226 *et seq.*
Christaller demand cone, 159
Cities in medieval Europe, 82 *et seq.*
City growth, 12
Closed orbit periodicity, 41
Cobweb theorems, 248
Co-dimension, 14
  infinite, 17
Co-rank, 14
Commuting mappings, 16
Comparative statics, 117
  of urban spatial structure, 93 *et seq.*
Compensation curves, 251
Competition, 166
  and modal choice, 212
  coefficients in ecology, 237
  for resources, equations, 49 *et seq.*, 150 *et seq.*, 236 *et seq.*, 239
Complex dynamic behaviour,
  and difference equations, 240
Complexity, 225
  and stability, 237
  in prey-predator equations, 265
Composite attractiveness factors, 105 *et seq.*, 191, 297
Conflict set, 21
Conflict states, 5
Constraint catastrophes, 30, 81, 142
Constraints, 6
Consumers' behaviour, 95, 175
Consumers' surplus, 172, 198, 271
Continuity, 246
Continuous space, 202
  SIA models, 174
Control manifold, 20, 26, 51
  for butterfly catastrophe, 77
Control theoretic models, 171 *et seq.*
Control variables, 2, 203
  in chemistry, 227
Convergence
  exponential, 39
  oscillating, 39
Cost curves for retail model, 119
Critical depensation, 251
Critical parameters, 5, 162
Critical points, 131
  and control theory, 171
Critical surface in parameter space, 128
Criticality, 64, 127, 138

Index

Critiques of catastrophe theory, 92
Cubic equation, algebra of, 25
Cubic functions, plots of, 19
Cusp catastrophe, 25 *et seq.*
    and Amson's fourth law, 72
    and bimodal distributions, 204
    and business cycles, 253 *et seq.*
    and central place theory, 75
    and city growth, 12
    and depressions, 255
    and genetic switching, 233
    and modal choice, 8 *et seq.*, 206 *et seq.*
    and property prices, 76
    and settlement evolution, 86
    and structural change, 82 *et seq.*
    and sudden urban grwoth, 89 *et seq.*
    and travelling frontier, 247
    and urban property prices, 76
    in chemistry, 227
Cusp, false, 30
Cusp surface, 4, 14
Cycles, business, 253 *et seq.*

Darwinism, 234
Degeneracy, degree of, 14
Degenerate singularities, 17
Delay conventions, 6, 20 *et seq.*
Demand cone, Christaller, 159
Densities, urban, 69 *et seq.*, 84, 99
Dependent variables, 2
Depensation curves, 251
Depressions, 253
Derivatives
    of disaggregated retail flows, 297 *et seq.*
    of retail flows, 295 *et seq.*
Deterrence functions, 106, 190
Development, 196 *et seq.*
    modelling, 143
    models, research on, 197 *et seq.*
    possible states, 140, 193
Developmental biology, 196, 232
Diffeomorphism, 16
Diffeotype, 16
Difference equations, 267
    and shopping centres, 260 *et seq.*
    for urban models, 114 *et seq.*
    for population dynamics, 240 *et seq.*
Differential equations, 267
    and bifurcation, 33 *et seq.*
    and catastrophe theory, 58 *et seq.*
    for retail model, 110 *et seq.*
    for population dynamics, 236

Differential topology, 1, 15
Differentiation, 246
Diffusion, 268
  processes, 61
  spatial, in biology, 238
Digraph method, 238
Disaggregated models, dynamical equations for, 116
Disaggregation
  of residential location model, 146 *et seq.*
  of retail model, 149 *et seq.*
  principles of, 101 *et seq.*
Discrete change, 1
Disequilibrium, 155 *et seq.*
Dissipative structures, 229 *et seq.*
Divergence, 5, 12, 74
Domino effect, 148
Dual catastrophes, 28
Dynamic modal choice models, 210 *et seq.*
Dynamical analysis
  equations for in urban models, 110 *et seq.*
Dynamical equations
  Lagrangian forms of, 291
Dynamical systems, 34 *et seq.*, 57 *et seq.*
Dynamics of urban spatial structure, 155 *et seq.*

Ecological analogy, 111, 142, 150 *et seq.*
Ecological models
  travelling waves in, 246
Ecological niche, 201, 235
Ecology, 236 *et seq.*
Economics, 248 *et seq.*
Economies of scale, 90, 160, 161, 175, 199
Elementary catastrophes, 14, 28 *et seq.*
Embedding, 62
  and mathematical programming, 108, 280
  dynamic, 38, 94
Embryology, 234
Employment, 101, 114
Entrepreneurial differential equations, 172
Entropy
  Kullback form, 288
Entropy maximising methods, 97, 105, 225, 271
Environmental gradient, 239, 246
Environmental quality, 99
Equilibrium, 3, 33
  far-from, state, 229
  thoermodynamic, 156
Equilibrium manifold, 92
Equilibrium point analysis
  for retail model, 116 *et seq.*
Equilibrium points, 37 *et seq.*, 39
  of Lotka-Volterra equations, 47
Equilibrium solutions, multiple, 5

Index

Equilibrium states, 14, 267
Equilibrium surface, 61
Evolution, 196 *et seq.*
   linear programming model of, 200
   of new structures, 199 *et seq.*
   of settlement patterns, 84 *et seq.*
   of urban structure, 191 *et seq.*
   paths of, 143
Evolutionary biology, 143, 196, 234 *et seq.*
Evolutionary model, 300
Explosive reactions, 227
Exponential convergence, 39
Exponential growth, 43

Facility size, benefits of, 79
False cusp, 30
Feedback, 45, 61, 112, 264, 268
Fields, interacting, 97
Firm, theory of, 175
Fisheries, management of, 249 *et seq.*
Flow-density curve, 219
Fluctuations, 55, 61, 155, 228, 268
Fast dynamics, 61, 268
Fast equations, 58
Fast foliation, 58
Fast returns, 253
Fold catastrophe, 22 *et seq.*
   and Amson's third law, 71
   and central place theory, 75
   and centre size, 80
   and fisheries yields, 252
   and Lotka-Volterra equations, 48
   and retail centre size, 129
   and spatial structure, 10
   and speed-flow relationships, 218 *et seq.*
Folds, 5
   in equilibrium surface, 3
Food nets, 237
Forecasting, conditional, 64, 192
Form, stability of, 1
Frontier, travelling, 246
Functions,
   families of, 3, 14, 18
   types of, 17

Geographical analysis, 6
Geography, historical, 84
General systems theory, 235
Genes, 233 *et seq.*
Genetic nets, 237
Genetic switching, 233
   urban analogues of, 256 *et seq.*
Gradient systems, 2

324

Graphical presentations, forms of, 34 *et seq.*
Grazing threshold, 238
Groups surplus, 283 *et seq.*
Growth
    catastrophic modes of, 69 *et seq.*
    equations, 43 *et seq.*, 111, 112
    of fish populations, 250
    sudden urban, 89 *et seq.*

Habit, 8, 12, 206
Hamiltonian, 248
Harvesting functions for fish population, 250
Health services, 98
Hessian matrices, 38
Hexagonal patterns, 177
Hierarchical structure, 162, 173, 176, 235, 268, 298
Higher order catastrophes, 28 *et seq.*
Historical accidents, 142
Historical analysis, 192 *et seq.*
Historical geography, 84
Homeostasis, 246
Host-parasite models, 245
Hotelling integrability conditions, 281
Housing, 88, 182
Hysteresis, 5, 8 *et seq.*, 12, 74
    and fish populations, 252
    and modal choice, 206 *et seq.*, 208 *et seq.*, 211
    empirical evidence for, 9, 210

Ideal gas law, 70
Imitation, and modal choice, 215
Imperfect gas laws, 70
Implicit functions, 131
Income levels, 12
Independent variables, 2
Inductive vs. deductive, 63 *et seq.*
Inertia, 115
Inhibition coefficients, 237
Input-output models, 189
Instability, structural, 49
Interacting chemicals in biology, 232
Interacting fields, 99
Interacting mixtures in chemistry, 226 *et seq.*
Interacting populations, 46 *et seq.*, 54 *et seq.*
Investment function, 254

Jumps, 5, 74
    interpretation of, 142

Kinetic equations, 168 *et seq.*
    in chemistry, 226 *et seq.*
Kullback entropy, 288

Index

Lag structure, 257
Lagrangian formulations, 112, 130, 291 *et seq.*
Land, residential, 182
Land use accounts, 100, 185
Lattice
   square, 162
   triangular, 161
Laws, urban, 69 *et seq.*
Levels of approach to dynamical systems theory, 57 *et seq.*
Limit cycles, 41
Limits to growth, 189
Linear difference equation model, 261
Linear programming, 200
Linked subsystems, 55
Logistic growth, 45, 113, 133, 158, 250
   difference equations for, 241 *et seq.*
   for interacting populations, 49, 54 *et seq.*
Lotka-Volterra equations, 46 *et seq.*, 150 *et seq.*
   and neural nets, 237
   and war, 249
   as difference equations, 245
   in biochemistry, 234
   in chemistry, 228 *et seq.*
   in ecology, 236 *et seq.*
   with diffusion, 238
Lowry model, 94, 99, 108, 116, 148, 179, 180, 185, 189, 285 *et seq.*

Macro scale, 12, 67 *et seq.*
Manifold, 14
   behaviour, 20, 35
   control, 20, 51
   control, for butterfly catastrophe, 77
   equilibrium, 92
   slow, 58
Market areas, 179
Markovian kinetic equations, 168 *et seq.*
Mathematical concepts, and catastrophe theory, 14 *et seq.*
Mathematical programming, 108 *et seq.*, 271 *et seq.*
Maxwell convention, 21, 28
Memory variables, 257
Meso scale, 10 *et seq.*, 93 *et seq.*, 155 *et seq.*
Micro scale, 8 *et seq.*, 203 *et seq.*
Micro-simulation models, 223
Mixed conventions, 22
Modal choice models, 8 *et seq.*, 206 *et seq.*, 210 *et seq.*
Model
   retail, differential equations for, 110 *et seq.*
   employment location, 114
   retail, equilibrium point analysis, 116 *et seq.*
   urban retail structure, 94 *et seq.*

Models
    approaches to, 60 *et seq.*
    mathematical programming foundations of, 108
    planning applications of, 64 *et seq.*
    spatial interaction, 95 *et seq.*
    SIA, 174 *et seq.*
    structural stability of, 15
    urban, 93 *et seq.*
Molecular biology, 234
Morphogenesis, 234
Morse singularities, 17
Multiple equilibria in economics, 248
Multiple minima, 20
Multiple solutions, 5, 157
Multiple states, 6, 171
Multispecies prey-predator model, 237
Mushroom catastrophe, 88

Net, randomly connected, 233
Neural nets, 237
Nicholson-Bailey host-parasite models, 245
Noise, 61, 268
Non-gradient systems, 33 *et seq.*
Nonlinearities, 45, 97, 106, 111, 118, 161, 204, 253
Normal factor, 12, 28

Order from fluctuations, 156 *et seq.*, 228 *et seq.*
Optimum size and location of facilities, 79, 93
Opulence, 69
Oscillatory behaviour, 36
Oscillatory convergence, 39

Parabolic umbilic catastrophe, 88
Parameter space and retail model, 128
Perfect delay convention, 21, 206
Periodic behaviour, 41
Physical chemistry, 226 *et seq.*
Phytoplankton, 238
Planning, 269
    and control, 192
    applications of models, 64 *et seq.*, 93
Population activities, 93 *et seq.*
Population dynamics
    difference equations, 240 *et seq.*
    differential equations, 236
Populations, interacting, 46 *et seq.*
    logistic growth of, 54 *et seq.*
Potential function
    and homeostasis, 247
    and social costs, 89
    and welfare, 82
    construction of, 68, 78

Index

Prey-predator equations, 46 *et seq.*, 150 *et seq.*
 as difference equations, 245
 in biochemistry, 234
 in chemistry, 228 *et seq.*
 in ecology, 236 *et seq.*
 complexity in, 265
Production functions, 190
 and evolution, 197
Profit maximising, 204
Programming, mathematical, 108, 271 *et seq.*
Property prices, 75 *et seq.*
 and cusp catastrophe, 76
Psychological factors, and modal choice, 215
Publicity, and modal choice, 215
Pucker point, 140

Qualitative vs. quantitative, 63 *et seq.*

Random effects, 157, 167
Random utility models, 105, 283 *et seq.*
Randomly connected food nets, 237
Rank, central place, 13
Recessions, 253
Recreation, 98
Repeatability, 246
Repellors, 33
Residential attractiveness terms, 106
Residential location model, 144 *et seq.*, 181, 285
 Boolean, 257 *et seq.*
 disaggregation of, 104
 dynamical equations for, 115 *et seq.*
Residential segregation, 168
Residential structure, 98
Residential supply, 182
Resilience, 193
Resource limits, 44
Resource management, 249 *et seq.*
Resources, competition for, 49 *et seq.*
Retail centres, evolution of, 199 *et seq.*
Retail centre size, 261
Retail model
 differential equations, 110 *et seq.*
 disaggregation of, 102
 equilibrium point analysis, 116 *et seq.*
Retail services, use of, 183
Retail structure, 94 *et seq.*
Retail supply, 184
Retailing (see also shopping)
 and structural evolution, 141
 flow-centre-size curves, 122, 124
Revenue curves, for retail model, 119
Revival of cities, 82 *et seq.*
Riemann-Hugoniot catastrophe, 70
Robustness, 193

Saccadic city, 74
Saddle points, 39
Savings functions, 254
Scale, 6, 60
Schools, 98
Secondary impulse, 148
Segregation, residential, 168
Selection process, 234
Self duals, 30
Separatrix, 39
Separatrix crossing, 142, 162
Settlement pattern evolution, 84 *et seq.*
Shopping (see also retailing), 8
Shopping centres
    and control theory, 172 *et seq.*
    and difference equations, 260 *et seq.*
    size, 78 *et seq.*
SIA models, 174 *et seq.*
    and central place theory, 179 *et seq.*
Simple systems, 6
Simulation models, 161
    micro, 223
Simultaneous criticality, 130
Singularities, 14, 18, 61, 117
    denegerate, 17
    Morse, 17
    types of, 4
Slow dynamic, 61, 268
Slow equations, 58
Slow manifold, 58
Slow returns, 253
Slums, 88 *et seq.*
Social costs, and potential function, 89
Space-time variables, 61
Spatial diffusion, 238
Spatial interaction, 10, 95 *et seq.*, 174
Spatial patterns, and hierarchical structure, 162
    in biology and chemistry, 233
Spatial structure, 10, 69 *et seq.*
    comparative statics of, 93 *et seq.*
    urban dynamics of, 155 *et seq.*
Speed-density curves, 219
Speed-flow relationships, and the fold catastrophe, 218 *et seq.*
Speeds of change, relative, 59 *et seq.*
Splitting factor, 12, 28
Square lattice, 162
Stability, 33, 64, 198, 237, 263
    and Lotka-Volterra equations, 47
    concepts of, 5
    of equilibrium points, 38
    of zones, 193
    structural, 19
    structural, of models, 19

Index

State space, 33
State variables, 2
States, multiple, 6
Static models, embedded in dynamic frameworks, 38 *et seq.*
Stochastic terms
    and kinetic model, 169
    in urban models, 157
Structural change, 82 *et seq.*
Structural instability, 49
Structural stability, 19
    of models, 15
Structure, urban, evolution of, 191 *et seq.*
Subscript lists, 202
Sudden urban growth, 89 *et seq.*
Suppliers' behaviour, 95
System description, 59
System types, 225

Taylor expansions, 18
Technological change, 90
Thermodynamic branch, 230
Thermodynamic equilibrium, 156, 227
Thom's theorem, 4, 18, 68, 92, 247
Thresholds, 12, 74, 190
    and habit, 209
    grazing, 238
Trajectories
    in state space, 33 *et seq.*
    on manifold, 14
    sampling of, 171
    types of, 39 *et seq.*
Trajectory sketching, 37 *et seq.*
Transformations
    examples of explicit, 73
    to canonical form, 5
Travel, disbenefits, 79
Travelling frontier, 246
Travelling waves, 234
    in ecological models, 246 *et seq.*
Triangular lattice, 161

Unfolding, 19
Urban growth, sudden, 89 *et seq.*
Urban laws, 69 *et seq.*
Urban models, 93 *et seq.*
    and Boolean algebra, 256 *et seq.*
Urban retail structure, 94 *et seq.*
Urban spatial structure, dynamics of, 155 *et seq.*
Urban structure, evolution of, 191 *et seq.*
Urban subsystems, 109
Urban systems, and other disciplines, 225 *et seq.*
Utilities, urban and rural, 89 *et seq.*
Utility, components of for facility size, 79

Utility function for driver behaviour, 219
Utility maximisation, 175, 204
    and modal choice, 207 *et seq.*

Verhulst logistic model, 158

van der Waal's equation, 70
War, mathematics of, 248
Wave, travelling, 234
    in ecological models, 246 *et seq.*
Weaver's classification of systems, 6
Workplace location, 98

Zeeman catastrophe machine, 7, 27
Zeeman's six steps for analysis, 267
Zooplankton, 238